Negotiating Personal Autonomy

Negotiating Personal Autonomy offers a detailed ethnographic examination of personal autonomy and social life in East Greenland.

Examining verbal and non-verbal communication in interpersonal encounters, Elixhauser argues that social life in the region is characterised by relationships based upon a particular care to respect other people's personal autonomy. Exploring this high valuation of personal autonomy, she asserts that a person in East Greenland is a highly permeable entity that is neither bounded by the body nor even necessarily human. In so doing, she also puts forward a new approach to the anthropological study of communication.

An important addition to the corpus of ethnographic literature about the people of East Greenland, Elixhauser's work will be of interest to scholars of the Arctic and the North, Greenland, social and cultural anthropology, and human geography. Her conclusion that, in East Greenland, the 'inner' self cannot be separated from the 'public' persona will also be of interest to scholars working on the self across the humanities and social sciences.

Sophie Elixhauser is an Honorary Research Fellow in the Department of Anthropology at the University of Aberdeen, UK. Her research interests include human–environmental relations and interpersonal communication in East Greenland, and the human dimensions of climate and environmental changes in the European Alps. She currently works in the field of migration in Munich, Germany.

Arctic Worlds
Communities, Political Ecology and Ways of Knowing
Series Editors:
David Anderson and Robert J Losey

This series aims to integrate research from across the circumpolar Arctic from across the humanities, social sciences, and history of science. This region – once exotised as a remote and unknown "blank spot" – is now acknowledged to be the homeland of a variety of indigenous nations, many of whom have won or are seeking home rule.

The Arctic was the central axis of frozen confrontation during the Cold War. At the start of the 21st century it is a resource hinterland offering supplies of petroleum and minerals for aggressively new markets with great cost and risk to the environment. The indigenous nations of the region are unique for their "ways of knowing" which approach animals and landscape as alive, sentient entities. Many share cultural commonalities across the Arctic Ocean, sketching out a human community that unites disparate continents.

This series takes history seriously by bringing together archaeological work on ancient Arctic societies with ethnohistorical studies of the alternate idioms by which time and meaning are understood by circumpolar peoples, as well as science and technology studies of how the region is perceived by various scientific communities.

Titles in series

Negotiating Personal Autonomy
Communication and Personhood in East Greenland
Sophie Elixhauser

For more information about this series, please visit www.routledge.com/Arctic-Worlds/book-series/ARCTICW

Negotiating Personal Autonomy

Communication and Personhood in East Greenland

Sophie Elixhauser

LONDON AND NEW YORK

First published 2018 by Routledge

2 Park Square, Milton Park, Abingdon, Oxfordshire OX14 4RN
52 Vanderbilt Avenue, New York, NY 10017

Routledge is an imprint of the Taylor & Francis Group, an informa business

First issued in paperback 2020

British Library Cataloguing-in-Publication Data

A catalogue record for this book is available from the British Library

Library of Congress Cataloging-in-Publication Data
A catalog record for this book has been requested.

ISBN: 978-1-138-06855-1 (hbk)
ISBN: 978-0-367-59222-6 (pbk)

Typeset in Times New Roman
by Apex CoVantage, LLC

Contents

Figures

Maps

Acknowledgements

Many years from now, this project started with a vague dream about Greenland, about this place so far up north, with its fascinating history, environment, and population. Slowly and step by step, I was able to combine this interest with my PhD studies in anthropology, which led to many months of fieldwork, subsequent visits and stays, and a connection with the Iivit and their country that I would not want to miss. Supported by numerous people, this dream ultimately led to this book. I want to express my deepest gratitude to all of the people who have helped me on this journey, and I am sorry that I cannot list everyone in person.

First and foremost, I wish to thank the families in Sermiligaaq who opened their homes to me, integrated me into their families, and let me become part of their everyday lives. Their friendship has been crucial to my work, and their care, patience, and humour made my stay in East Greenland a time I will always enjoy remembering. I thank the family of Nora and Joel Nathanielsen, the family of Charlotte and Emanuel Boassen, and the extended family of Pavia Nathanielsen. In 2008, two of my close friends in Sermiligaaq passed away, both of whom helped and taught me a lot. I am very sorry that Basiba Nathanielsen and Katrine Singertat are no longer with us. I also thank the families Nina and Viggo Kalia, Kista and Simujoq Nathanielsen, Margrethe and Levi Uitsatikitseq, and Julie Nathanielsen for their friendship and hospitality. I am grateful to Johan Uitsatikitseq for his support of and interest in my work, and for letting me use his desk in the school building during evenings and weekends.

In Tasiilaq, I am grateful to Robert Peroni from the Red House, who made my first three-months stay in Tasiilaq possible in 2005. In 2006–7, Robert once and again let me stay at his private house in Tasiilaq, and he has put me in touch with the first people whom I got to know in Sermiligaaq. Help and support also came from Inge Weber, Peter Heimburger, and Tobias Ignatiussen. Many thanks to my friends in Tasiilaq for their hospitality and friendship: Dorthe Ale and her extended family; Asta Mikaelsen; her father Knud, who passed away in 2008; Holger Amelung; and Dorthea and Kenneth Liedtke – Dorthea sadly left this world in 2014. Thanks to Laura Emdal Navne for her friendship and the emotional support throughout the last months of my fieldwork, and to her and her husband Johan for their hospitality. I thank Eva Mätzler for her kind hospitality in Ittoqortoormiit. Prof. Yvon Csonka invited me to present my work at Ilisimatusarfik, University of Greenland, and supported my stay in the Greenlandic capital. Steen Yeppson has been a great guide in Nuuk, and has explained to me a lot about the city and its people.

Prof. Tim Ingold at the University of Aberdeen in Scotland has been the supervisor of my East Greenland project that ultimately led to this book. This book owes a great deal to his manifold comments, suggestions, and encouragements since the very start of my East Greenland project in 2004 up until its completion. It was a great pleasure to work with Tim Ingold, and several of the theoretical key strands in this book developed in inspiring discussion with him. Many thanks! Dr. Nancy Wachowich was the second supervisor of my East Greenland project. She provided emotional support throughout my fieldwork in East Greenland, and her advice – based on much experience with and sensitivity towards the Inuit – helped me a lot. She has supported my research on various levels, in coming to terms with my fieldwork material, finding the right conceptual approaches, and so forth. Prof. Dr. Frank Heidemann at LMU Munich supported the preparations for my project prior to my enrolment at the University of Aberdeen. He provided me institutional affiliation for the first year of my PhD and was supportive in Munich after I returned from Scotland to Munich during the writing period.

Funding for this project came from the College of Arts and Social Sciences, University of Aberdeen (6th Century Studentship), the University of Aberdeen Principal's Small Grants Fund, and the Department of Anthropology, University of Aberdeen. The financial support provided by these organisations is greatly appreciated.

Moreover, I would like to thank Anni Seitz for the many fruitful discussions in the course of our joint documentary project and for sharing my Sermiligaaq (and Tasiilaq) life for some weeks in the frame of the project. I am thankful to Claudia Regel for her continued interest in, and advice on, my experiences in East Greenland. My sincere thanks go to my colleagues and friends in Aberdeen, Munich, and elsewhere; for their friendship, support, and inspiration; and for their comments on draft chapters of my thesis: Stefanie Belharte, Sabine Deiringer, Katy Fox, Janne Flora, Caroline Gatt, Gabriel Klaeger, Franz Krause, Anna Kugler, Amber Lincoln, Swenja Greska, Claudia Regel, and Susanne Schmitt. Thanks to the members of the Writing Up Seminar at the University of Aberdeen and the Anthrodocs at the LMU Munich.

Finally, I am grateful to my family for their continued support. Special thanks to Ilse Hartgasser for her help and interest (sadly she passed away in 2012), to my brother Sebastian, and to my mother Ursula, who has been a source of emotional and practical support throughout. Last but not least, I thank Martl Jung for being there for me and for becoming a part of my Greenland world.

Introduction

One morning in 2006 in the village of Sermiligaaq, I was asking my friend Maline if her daughter Thala would be going to the kindergarten that day. Maline answered, '*nalivarnga . . . nammeermii*' ('I don't know . . . she has to decide herself'), and passed the question on to Thala. The three-year-old girl raised her eyebrows, showing her consent. I was quite surprised that such a decision was up to the girl and not to be decided by her mother. In the course of my ethnographic fieldwork in the Ammassalik region, East Greenland, conducted between 2005 and 2008, added by subsequent visits up to now, I came across many similar examples which pointed to the high value of personal autonomy in children and adults alike. Tunumeeq (the inhabitants of Tunu, i.e. East Greenland), once again, have used the expression *nammeq* (it's up to you, decide by yourself), and different word forms of it such as *nammeermii*, to indicate when particular decisions were up to a person and not to be influenced by other people. Apart from particular words and ways of speaking, this autonomy was further expressed through facial expressions, bodily postures, or communicative silences.

Sermiligaaq, the main focus of my ethnographic field research, is a village on the east coast of Greenland with 197 inhabitants in 2016 (Kommuneqarfik Sermersooq 2016). It is the northernmost settlement in the Ammassalik region located just south of the Arctic Circle. Apart from the small regional capital Tasiilaq, which comprises of 2010 inhabitants (Statistics Greenland 2017), the region has five villages (including Sermiligaaq), with population numbers ranging from ca. 100 to 300 people. The people identify as Iivit, which means Inuit, i.e. people, in the East Greenlandic language; they also call themselves Tunumeeq (East Greenlanders) or sometimes Kalaalleq, which means Greenlanders more generally. Until approximately the middle of the 20th century, the formerly semi-nomadic Inuit population made their living from hunting sea mammals, supplemented by some fishing and gathering of plants. Today, only a small percentage of the Iivit still earn their livelihood from hunting and fishing. Wage labour for the local administration and some private companies provides the main source of income. Tasiilaq nowadays comprises a Danish population of around 10 percent (including some other nationalities), yet in the villages one encounters only few Danes. In this book I mainly focus on the Iivit, and I will specify when speaking of East Greenlanders with a Danish (or foreign) background.

I visited the region for the first time in spring/summer of 2005, during which I volunteered for a guest house located in Tasiilaq, being able to start with some research and language studies on the side. The summer of the following year (2006), I returned for my main one-year period of ethnographic fieldwork and shifted my research focus and domicile from Tasiilaq to the village of Sermiligaaq. I established close friendships with Sermilingaarmeeq (inhabitants of Sermiligaaq) and other Tunumeeq, and with two extended families in particular with whom I shared day-to-day life during my time in the village. For most of my time in Sermiligaaq I lived with the family of my friend Maline, her husband, and their five children.[1]

My role as a language apprentice throughout my fieldwork played a major part in the development of my research topic on everyday communication and personal autonomy, and on how people negotiate the tension between personal autonomy and social responsibility. Through being immersed in Tunumiusut, the East Greenlandic language, and communicating in this language right from the start (few of my friends and informants speak English, and I speak only little Danish), over the months I was able to understand more and more of what people said to me and to each other. This went along with a growing awareness of the many nonverbal modes of communication that feature so prominently among the Iivit. My study of the language and my endeavour to learn as many details as possible about the people's way of life repeatedly drew my attention to the high value of people's personal autonomy – expressed through the concept of *nammeq* – which is closely linked to a particular understanding of personhood. Contrary to a view of autonomy in the sense of complete individual freedom, however, this kind of autonomy is not tantamount to a complete independence from others. This is similarly reported from various other small-scale societies around the world, many of which are subsumed under the label 'hunter-gatherers' (e.g., Biesele 2004; Bird-David 1987; Gardner 1991; Myers 1986; Ridington 1988; Sugawara 2004). This autonomy is closely embedded in relationships with others and refers to a kind of knowledge which is not given in advance but which a person needs to find out for him- or herself by being observant and attentive. An autonomous person, accordingly, is not conceived of as a bounded, indivisible entity, but as constituted through attentive relations with others.

In this book, I take up a communicational perspective on personal autonomy. I explore various modes of communication, including ways of speaking and not-speaking, facial expressions, gestures and bodily movements, and how people attune their communicational practices to those of other people. The examples used to illustrate this broader theme are drawn mainly from situations of interpersonal communication among inhabitants – as observed, participated in, and learned by myself – and to some extent also encompasses communication with non-human persons and animals. By regarding communication as an embodied practice that involves persons constantly on the move – be it through literally changing place or through a gesture – this book aims to show that we can understand communication only when we consider the broader themes of personhood, perceptions of self and others, embodiment, and movement.

The main questions that arose in the course of working on my East Greenland material include the following: What types of communication are compatible with autonomous personhood? How is the high value placed on personal autonomy reflected in different understandings of the person? By addressing these questions, this book seeks to contribute to discussions among ethnographers working in Inuit, Yupiit, and Inupiat communities across the circumpolar North about the importance of respecting other people's personal autonomy in interpersonal encounters. Questions of autonomy have been analysed in relation to Inuit in Canada (e.g., Briggs 1970, 1998; Madden 1997), and Yupiit and Inupiat in Alaska (e.g., Fienup-Riordan 1986; Morrow 1990, 1995, 1996), but not in the context of Greenlandic Inuit. Indeed, few East Greenland ethnographies address the notion of personal autonomy, apart from some short mention once in a while (e.g., Gessain 1969: ch. 5; Hovelsrud-Broda 2000a: 154; Thalbitzer 1941: 625). It is much the same with (ethnographic) studies from West Greenland (but see Flora 2012). In stressing the link between ways of communication, the importance of the value of personal autonomy and a particular understanding of personhood, this book aims to contribute to a more nuanced understanding of East Greenlanders' ways of communicating between themselves and in their encounters with others.

Fieldwork context and how this book's theme developed

My research in East Greenland consisted of a main, one-year period of fieldwork, and two shorter fieldtrips with a duration of three months and one month respectively, summing up to fifteen months in total. Since that time I have returned to the region for visits every few years. I first became interested in the Greenlandic east coast during a working holiday in the community of Aassiat, West Greenland, in summer 2004, when I met some people who had previously been there. After enrolling in the PhD programme in social anthropology in autumn 2004 – for an initial year in Munich before changing to the University of Aberdeen in autumn 2005 – I spent three months in East Greenland in spring 2005. During that time, I volunteered at the Hotel The Red House, a guesthouse in Tasiilaq, the urban centre of East Greenland. I built up my first contacts, and started to learn Tunumiusut, the East Greenlandic language, which became the main means of communication throughout my fieldwork. Since it was tourist offseason, there was enough time to explore the region, and I visited the village of Sermiligaaq for the first time in May 2005. I decided to move the major focus of my fieldwork to Sermiligaaq when I returned for my one-year period of fieldwork the following summer. Returning to East Greenland in July 2006, I initially kept my main base in Tasiilaq. In the course of the first few months, I spent more and more time in Sermiligaaq, where I soon established friendships. After initially visiting at a friend's place in Sermiligaaq for some shorter periods of time, and staying for some weeks by myself in a small hut (called *paraki*) in the centre of the village sometimes used by construction workers, in December 2006 I moved in with my friend Maline's family, who had invited me to stay with them. I had met Maline at the various meeting places in Sermiligaaq during my initial weeks in the village. I continued to live with her,

her husband, and their five children – at that time ranging from one and a half to twelve years in age – throughout the rest of my fieldwork until summer 2007, whilst still sometimes spending shorter periods of time in Tasiilaq. Apart from my host family and their relatives, I maintained close contacts with another extended family in Sermiligaaq. This family group encompasses several households in Sermiligaaq that I visited on a daily basis, sometimes also staying overnight, and on occasion I joined family members for hunting and fishing trips (they are mostly hunters, whereas my host family draws its income from wage labour). In Tasiilaq, I spent much time with Sermilingaarmeeq living in (or visiting) town, and tried to become familiar with the networks among and around them. Apart from Sermiligaaq and Tasiilaq, I visited all the other villages in the Ammassalik region, and I spent a few weeks in Ittoqqortoormiit in Northeast Greenland, and in Nuuk, West Greenland.

In order to understand the background of my research topic and my theoretical approach, it is important to mention that I did not plan at the outset to investigate communication and personal autonomy, and this research topic only developed in the course of, and subsequent to, my fieldwork. I began my project with a sustained interest in human-environmental relations, which led me to focus my initial research proposals on East Greenlanders' relationships with their environment. I intended to investigate inhabitants' perceptions of particular landscape features, such as glaciers, mountains, and weather and light conditions, and their practices of orienteering and wayfinding. These interests rested upon the premise that when speaking about the environment one has to acknowledge the positioned involvement of the people – and also of the ethnographer – as participants within the environment in question. My project drew (and draws) from Tim Ingold's (2000) ecological and relational approach, which presupposes that persons and organisms are ontologically equivalent, thereby bridging the much criticised opposition between 'nature' and 'culture' (on the latter, see e.g. Descola and Pálsson 1996; Hornborg and Pálsson 1996; Roepstorff et al. 2004). The human person is constituted within a nexus of relations that embrace both other humans and non-humans alike (Lave 1988). It is considered 'not as a composite entity made up of separable but complementary parts, such as body, mind and culture, but as a singular locus of creative growth within a continually unfolding field of relationships' (Ingold 2000: 4–5). Personhood, moreover, as has been demonstrated by ethnographic research in different parts of the world, is not generally limited to human beings (Harvey 2005; Ingold 2000: ch. 6; Milton 2002). Such understanding, which has often been framed in terms of so-called animism – and the close intertwining of the human and non-human spheres that it builds upon – has been found among the Inuit, in East Greenland, and across the circumpolar North (e.g., Dahl 2000; Fienup-Riordan 1986; Ouellette 2002; Thalbitzer 1930). The premise that, in East Greenland, one cannot neatly separate a human from a non-human realm, lays the foundation for my project.

My initial plans to study environmental perception, knowledge of certain landscape features, and the role of the different senses in orientation and wayfinding was influenced by my experiences in East Greenland during the first preliminary three-month fieldtrip in 2005. During that time, I had been attached to the Red House project, and together with another female volunteer, a foreigner like me, I had regularly been invited by some Greenlandic staff members to join in boat trips – in order to go hunting and fishing – some of which had lasted for several days. In this way, I had been able to get initial insights into matters concerning orientation out on the sea, knowledge about the weather, and so forth. Returning in July 2006, however, the situation was quite different. I was not working for the Red House any more, and I was accessing a new field site in Sermiligaaq. Though still able to sometimes join East Greenlandic friends on boat trips and with fishing and hunting activities, especially during the summer, I began to realise that my former research ideas had been very much directed towards the male realm in the society, which in the long run is not easy for a woman to regularly participate in. The more I adapted to local ways of life, and the more I understood the language, the more I found I also had to adapt to local conventions with regard to gender (cf. Guemple 1995 on gender among Canadian Inuit). As a young single woman at the time, I felt that I was not supposed to join in men's subsistence activities, not at least without additional female company, for to have done so would have suggested interest in one of the men. Though I did not encounter any explicit rules in this respect, teasing and gossip would quickly indicate what was considered 'right' conduct.

Many other female arctic ethnographers, such as Jean Briggs, Barbara Bodenhorn, Cunera Buijs, and Nancy Wachowich, just to name a few, also experienced the effect of gender on one's fieldwork. Similar to my experience, Briggs once mentioned in an interview that being a woman did influence her research topic and that, for instance, 'As a man, [she] would not have been able to stay home with the women and children without making the men extremely anxious' (2005: 13, brackets added). All of the ethnographers mentioned above have studied themes that relate to the female, or shared female-male, realm of society, and their ethnographies deal with topics such as emotional life and the education of children (Briggs 1970, 1998), clothing and questions of identity (Buijs 2004), kinship and sharing practices (e.g., Bodenhorn 1993, 2000a, 2000b), or the life histories of Inuit women (Wachowich 2001). Similarly, my gender has strongly influenced my everyday life in the field, my research activities, and the kinds of people I have spent time with. My closest friends are thus all female, and though I have interacted with men as well, especially within the families I have lived with, and though sometimes I was able to join in the men's hunting world a little, my research represents the female side more intensively.

Hence, parallel to my attempts to find a role and activities that were approved of by other community members, and in addition to my continuing interest in way-finding/orientation and environmental knowledge, I became increasingly interested in how communication works so differently from what I was used to. Driven by my endeavour to learn the East Greenlandic language, and the many ways of communicating without words, I was repeatedly struck by communicational conventions and expectations that differed from those to which I was accustomed. In the course of time, I paid more and more attention to these communicational patterns and began to re-adjust my research questions accordingly. Apart from the specific words used, the broader contexts of speaking and not-speaking, particular gestures, and bodily postures caught my attention. My original emphasis on environmental perceptions thus shifted towards an increased focus on interpersonal communication, especially with regard to family relations, to town-village exchange, and to questions of how and by what means people communicate directly and over a distance. My research into East Greenlanders' ways of communicating recurrently led me to experience situations which pointed to the high value of personal autonomy and a particular understanding of the person. However, I did not initially frame these observations in terms of 'autonomy' and 'personhood', and only when I came to analyse my fieldwork material upon my return to Scotland, and later to Germany, did I begin to fully appreciate the importance of these concepts for my study. Hence, some of the theoretical key strands of this book have found their way into my research only after my fieldwork while putting my experiences into context.

Taking into account both the dissolution of the nature/culture dualism and the corpus of ethnographic literature on Inuit communities which has shown that 'there is a resonance between Inuit attitudes toward people and their attitudes towards the physical environment' (Briggs 1991: 259), it becomes clear that the different thematic foci of my project – human-environmental relations on the one hand and interpersonal communication on the other – do not actually lie very far apart. Using a broad definition of 'the environment', which embraces both human and non-human surroundings, communicative relationships among human beings are as much part of human-environmental relations as, for example, people's relations with a particular mountain or glacier. Moreover, taking into account the close involvement of human and non-human realms in East Greenland, the scope of interpersonal communication may even include people's communicative encounters with non-human persons such as animals, or certain half human, half non-human creatures. I therefore argue that in order to understand communicative practices among the Iivit of East Greenland, one needs to adopt a perspective which considers people's wider relations with – and within – a field of manifold human and non-human beings and entities.

Methodological approach

My linguistic apprenticeship throughout the time of my fieldwork has played a major role in me coming to appreciate different forms of verbal communication as well as situations in which talking is not appropriate. Apart from my endeavour to learn the specific meanings of the words uttered, my attention was drawn to the contexts and constellations in which conversations took place, to 'who' talked about 'whom' or 'what', and to the localities of different types of conversation as well as temporal issues. My language learning efforts were further closely linked to trying to understand the many communicational modalities apart from speech, such as gestures, facial expression, and ways of looking. Listening and asking about words and meanings, trying to understand what is going on, and, as it often appeared to me, to solve hundreds of 'communicational puzzles', formed a prominent part of my everyday life. Often when visiting people, when spending time at some communal meeting place, or when participating in different daily activities, I carried my vocabulary book with me and wrote down words, or asked about them. Villagers often enjoyed flicking through my booklet, whilst commenting upon my learning progress, and wrote down words for me or corrected my spelling. In addition, I also made extensive use of the limited available linguistic material (Gessain et al. 1986; Robbe and Dorais 1986),[2] and sometimes East Greenlandic friends would translate for me through finding easier words.

Since I speak only a little Danish (though I am able to read it and understand some of it), right from the start I communicated with East Greenlanders in Tunumiusut. This endeavour involved long periods of being in the company of East Greenlanders and not (quite) understanding what was being talked about, as well as recurrent difficulties in making myself understood, especially during the initial months of my fieldwork. All in all, however, being immersed into this language right from the start proved very beneficial for my language skills and speeded up my learning process. This was compounded by the fact that in Sermiligaaq I was usually the only foreign person (apart from occasional tourists in the summer), with no Danish resident living in the village at the time.[3]

For many villagers, my role as a language student provided a comprehensible reason for my prolonged presence in the community, whereas when trying to explain my research objectives, inhabitants often appeared not greatly interested in hearing about them (or found my explanations somewhat alien). More important in terms of being accepted as a long-term guest, or temporary inhabitant, seemed to be whether I was nice company, whether one could laugh with (and, of course, about) me, whether I was willing to adapt to local ways of life, to not impose myself on others, and to join in and help with any activity. My continuing interest and eagerness to learn from the people, and the fact that I spent so much time with people in Sermiligaaq and also with some people in Tasiilaq, was often positively commented upon, and I think that people came to trust that I would use the knowledge I acquired in responsible ways (or at least try my very best to do so). I decided to use pseudonyms throughout this book, apart from the names of some key political figures who have spoken to me in their professional capacity.

Apart from studying the language and other means of communication, my methodological approach was based on participant observation – that is, trying to learn and participate in as many arenas of everyday life as possible, such as the education of children, work in and around the house, berry picking activities or fishing and hunting trips, travelling with East Greenlandic friends to other settlements in the region and meeting their relatives, or attending various village events. Furthermore, I had many informal conversations on themes connected to my research, and I have conducted a number of semi-structured interviews, mostly with officials, for example, local politicians and people in different institutional settings. Some of these interviews were part of a documentary film project which I carried out together with a German friend and colleague.

Interviewing did not serve as a major method, as many inhabitants did not seem to appreciate the typical interview situation or being questioned out of context. I

Figure 0.1 The author with some friends at Sermiligaaq harbour, September 2006
(Photo: Anni Seitz)

found that most people were much more open to talk about issues connected to activities of the moment, their immediate surroundings, or particular contexts. Accordingly, so-called 'go-alongs' turned out to be an important research tool, which involved asking questions, listening, observing, and being attentive whilst accompanying individuals moving through and interacting with their human and non-human environment (Kusenbach 2003). If I was interested in specific issues, I would sometimes 'drop in' comments, hoping that my friends and informants would pick them up and voluntarily elaborate upon them. At other times I would wait for which topics would draw the interest of my companions, without influencing communication through questioning, thereby taking into account (and learning about) people's autonomy in deciding what and when to communicate.

I also used film and photography to learn about East Greenlanders' visual and aural perceptions, and modes of communication with and about other people or environmental features. I made extensive use of my digital camera, and I have frequently shown and discussed pictures with friends in order to elicit their perceptions and associations. Furthermore, together with my friend Anni Seitz, who was a student at the film academy Baden-Württemberg, Germany at the time, I made a documentary film, a village and family portrait, which pictures the daily life of the family I had been living with whilst taking into account various village activities and events and broader social and environmental contexts (Seitz and Elixhauser 2008). Anni came to Greenland twice during my one-year period of fieldwork, in late summer 2006 and in late winter 2007, staying for a few weeks each time (for some reflections on the filming processes, see Elixhauser 2009a). In connection with this documentary, apart from the purposes of ethnographic documentation, film elicitation turned out to be an interesting method. Through Sermiligaarmeeq's reactions when looking together at our film material, as with the photos I took, I learned about visual (and other sensory) perceptions, such as attentiveness and interest towards specific landscape features, other people, events, and so forth. Details people had been telling me, pointing out, or laughing about, and observations later on incorporated into conversations, jokes, and so on, taught me about important issues concerning interpersonal communication.

I am aware that this book presents a highly personal story, and that my research 'data' have been greatly influenced by my own presence, as is inevitable in any anthropological research. Accordingly, when speaking about East Greenlandic ways of life I refer to my observations and experiences whilst staying and living in East Greenland for prolonged periods of time between 2005 and 2008, and during subsequent stays, augmented by my continuing contact with friends via mail, telephone, and the Internet. Not only are my 'data' influenced by the particular times and settings of my research, and by my own background as a German female researcher, having spent some years in Scotland and having travelled to many places in this world, but also they are rich in information acquired through particular key informants – that is, people I am particularly close to and who have taught me a great deal about East Greenlandic ways of being. Although some particular people appear in this book more often than others, I have tried to take into account as many different perspectives and groups of people as possible, in different localities, i.e. the villages of the district and in Tasiilaq, in Ittoqqortoormiit in northeast Greenland, as well as a few places in West Greenland. In this way – and by having embedded my research, and the experiences it builds upon, within a corpus of comparative ethnographic (and other kinds of) literature about East Greenland, Inuit groups around the circumpolar North, and even people of other regions – I believe that some of my research outcomes do indicate tendencies that might well apply more generally.

Such a claim nevertheless points to an inherent dilemma of writing a book about Inuit society. The academic endeavour implies – or, one could say that due to academic conventions, this book necessitates – that one takes an authoritative stance, which often suggests some kind of certainty with regard to the interpretations taken, which, however, stands in sharp contrast to Inuit and other arctic native people's reluctance to claim any general validity for their own statements (cf. Morrow 1990, with respect to Yupik society). Being aware of this dilemma, I have tried my best to describe the processes of how I have come to suggest certain conclusions, by giving detailed accounts of various examples, their contexts, and people involved, including myself. I still regard myself as learning about East Greenland, and I will always be so; this book, thus, can only provide a contextually situated account of what I have so far learned about East Greenlandic people, their society, and culture since I first went there in 2005.

Chapter outline

To develop the argument of this book, each of the chapters of this book takes as its frame an arena, theme, or social setting from the inhabitants' everyday lives. This layout is not meant to reflect any type of social or spatial structure of East Greenlanders' lifeworlds.

Chapter One gives an overview of the main theoretical strands of this book. I discuss research on personal autonomy in so-called hunting and gathering societies and terms used to describe this autonomy such as 'independence', 'equality', and 'egalitarianism'. I introduce anthropological discussions on personhood and self, and give a brief introduction to anthropological studies of verbal and nonverbal communication and literature about so-called indirect communication. Communication, I argue, is a relational phenomenon which, apart from the communicating parties and other human actors, involves the broader contexts and settings of the communicative encounter. Communication regarded as 'joint attunement' and an embodied practice involves the constantly moving body. In Chapter Two I provide background information on the geographical and socio-cultural setting. I first introduce East Greenland, its history, and the ethnographic research that has been conducted there as well as the localities where my research took place. Arguing that interpersonal communication most fundamentally rests upon particular relations between self and others, I then briefly outline how human beings in East Greenland are understood cosmologically, and describe some important ways in which the Iivit relate to other inhabitants, including naming practices, kinship, and adoption.

In Chapter Three I move on to investigate people's everyday movements and processes of travel throughout the region, considering that the Iivit have always been, and to some extent still are, a highly mobile people. Presenting movement as intrinsic to communication, I explore how various co-travellers attune their movements to the movements of other persons, be they human or animal. Parallels are drawn between the ways that humans interact with animals, and how they interact amongst each other. I illustrate the dense social networks of people moving about in the sea and fjords, and the importance of particular places for people's encounters. Moreover, I describe practices of wayfinding and orientation, and the situational type of leader- and followership which emerges in the context of wayfinding and hunting. These examples show a tendency towards nonverbal forms of communication, such as gestures, mimicry, bodily movements, and communicative silences, and highlight the power of focal gaze and the importance of peripheral vision for communication. Verbal communication often takes indirect forms, be it by means of humour or via media such as the inter-boat radio, which make possible a range of communicational ambiguities.

Both for people frequently travelling and, to a greater extent, for those bound to the settlements, family life provides the indispensable backbone of social life. Accordingly, Chapter Four explores domestic life and communication among family members and cohabitants. The focus here is on interpersonal communication among human persons living in close proximity, and on the complex interplay between various social responsibilities and the respect for people's personal autonomy characterising these encounters. This chapter shows that the spatiality of the house influences people's communicational patterns – be it through the arrangements of rooms, walls, and doors, or items that provide a context for communicative encounters – and how houses are performed and appropriated by inhabitants. I consider the effects of the transition from the habitation of turf and stone houses to the use of the modern type of houses today, for instance, with regard to particular notions of privacy, practices of curiosity, and not paying attention to others. I take a closer look at the education of children, people's expectations about giving information, and the power of words which is very pronounced in East Greenland, not only in the context of family life.

In the fifth chapter, I broaden my scope beyond family life and the home by analysing the flows of visitors and guests entering and leaving a house, and the material goods they may bring along in order to be shared and redistributed. The practice of visiting and of being visited is reliant on various implicit verbal and nonverbal ways of showing or demanding hospitality, with respect to inviting, greeting, and receiving visitors and guests, and the subsequent exchange of stories and news. Material items such as foodstuffs and presents form an indispensable part of hospitality, in relation to both everyday visits and festivals such as Christmas or *milaartut*. Here, communication takes place by way of these goods, which may also be regarded as extensions of persons, and this type of communication is indeed quite explicit and visible. I show that different types of goods, such as Greenlandic country food and store-bought products, are treated differently with respect to sharing. On occasion, money also circulates between visitors/guests and hosts, which I discuss with respect to gambling.

I end Chapter Five with a question about what happens when people do not follow societal obligations and expectations concerning sharing, or other values and ways of doing. This is addressed in Chapter Six, in which I examine informal ways of sanctioning deviant behaviour. Various indirect ways of speaking and notspeaking, ways of looking, and bodily postures may all indicate values and expectations concerning proper behaviour. I first focus on song and drum duels, which were a prominent means of dealing with conflicts in the olden days, arguing that some of their characteristics still feature prominently in people's contemporary daily communicative practices. I then discuss a number of informal communicative social sanctions, which build upon particular conventions of approaching and keeping informed about others such as ridicule and teasing, gossip, silence, and withdrawal. In the literature, informal social sanctions are often subsumed under the concept of 'public opinion'. There are difficulties, however, in extending this concept to the East Greenland case, partly because words are not detachable but are, rather, extensions of a person. Likewise, communication does not build upon a fundamental distinction between public and private themes of conversation, nor does the East Greenlandic understanding of personhood separate and then juxtapose an 'inner' self and a public person.

Chapter Seven broadens the view of interpersonal communication, which in previous chapters is mainly confined to encounters among humans, to include inhabitants' encounters with non-human beings as well. I will highlight the fact that a person in East Greenland may be human or non-human; non-human as well as human persons can have an impact upon Iivit's everyday lives and influence their communicative practices. For means of protection against malevolent beings, for staying in good health, and securing the integrity of one's own person, there are various behavioural guidelines, some of which influence the ways humans communicate amongst each other. But as I shall show, some characteristics of communication among humans, such as the power of words or likewise gaze, also apply to people's communicative encounters with non-human beings, as for instance regarding the figure of the *qivitteq*. The examples presented are symptomatic of an 'open' concept of the person, which calls for a particular carefulness in communication.

I conclude this book by summarising the key issues presented and by sketching some broader implications of my findings for the contemporary societal context in East Greenland. Touching upon themes such as the communication between Iivit and Danish residents and a number of social problems, I call for a need to better integrate the value of protecting other people's personal autonomy and integrity into the workings of the social, political, and economic institutions that currently govern a 'modern' East Greenland.

Notes

1 The majority of my fieldwork took place in the village of Sermiligaaq, East Greenland, supplemented by several months spent in the small regional capital Tasiilaq and visits to all the other villages in the Ammassalik region and Ittoqqortoormiit (as well as a number of towns in West Greenland). While many of my observations and analyses pertain directly to village life in Sermiligaaq and parts of the Tasiilaq population, some can be generalised to the East Greenlandic Inuit more broadly. To flag this distinction in the breadth and scope of application, I use the different group designations: 'Sermilingaarmeeq', 'Tunumeeq', etc.

2 Nicole Tersis's (2008) new East Greenlandic dictionary was published just a year after my main period of fieldwork.

3 During the winter of 2006–7, a young German construction worker temporarily lived with the family of his Sermiligaaq girlfriend, whom, however, I have only seldom met in the village. Throughout the last decades, a number of times Danish teachers have taken up residency in Sermiligaaq, such as happened shortly after I left Sermiligaaq in summer 2007. These teachers usually stay for periods of one or two years before returning to Denmark or elsewhere.

1 Setting the scene
Communication, autonomy, and personhood

The relationality of personal autonomy

The great importance of personal autonomy has been demonstrated for many so-called hunting and gathering societies around the world, and ethnographers have used different terms to describe this autonomy, ranging from 'individual independence', 'freedom of choice', and 'equality' to 'egalitarianism' (e.g., Gardner 1991; Myers 1986; Ridington 1988; Woodburn 1982). Speaking of 'hunter-gatherers', it is important to mention that I do not regard this expression as an economic (or socio-political) category, but as signifying a particular relationship between humans and their human, as well as non-human, environment.[1] As the 1965 symposium on band societies showed, many hunter-gatherers do not recognise formal authority, are non-competitive, and enjoy a great deal of personal and group autonomy (Lee and DeVore 1968). Numerous ethnographic accounts from around the circumpolar North highlight the value placed on personal autonomy and portray a complex interplay between social responsibility and group cohesion on the one hand, and personal autonomy on the other (e.g., Briggs 1970, 2001; Morrow 1990, 1996; Riches 2004; Sonne 2003; Therrien 2008).

Autonomy has often been associated with equality, a term which has been subject to much discussion (Helliwell 1995; Ingold 1986: ch. 9). Christine Helliwell (1995) criticises the conflation of autonomy and equality in many anthropological writings, and argues that one must distinguish between an 'equality of condition' and an 'equality of opportunity', of which only the latter implies a strong emphasis on autonomy (Helliwell 1995: 360). She stresses that societies in which people enjoy a great deal of autonomy may indeed reveal inequalities between one individual and another, yet it is important to differentiate 'between inequality as the outcome of autonomous group or individual action and inequality as socially imposed and as thus constituted through the denial of autonomy' (*ibid*: 361). The distinction to which Helliwell refers has often been proposed in the Western[2] tradition. It is, for example, what distinguishes liberalism (equality of opportunity – free-market competition) from socialism (equality of condition – the classless society). Ingold, however, has argued that the nature of equality in hunter-gatherer societies differs from both of these. The Western liberal argument is that differences of individual ability, given equal conditions, lead to inequalities of outcome. For hunters and gatherers, he explains that although differences of ability may be recognised and even celebrated, they are of no consequence, since the equality lies in the relationship of part to whole. Nor does this 'equality of relations founded in their commitment to the whole', again, correspond to the imposed equality of conditions of the socialist mode (Ingold 1986: 238; cf. Endicott 1999, on gender relations among hunter-gatherers).

Several ethnographers have observed that the notion of autonomy is deeply embedded in western experiences of sociality and personhood, which makes the translation of non-Western conceptions in terms of 'autonomy' rather difficult (Helliwell 1995: 362; cf. Myers 1988: 27; Ingold 1986: ch. 9). The autonomy of hunter-gatherers has often been confused with a common understanding of individualism of the kind, for example, proposed in the classical work of Louis Dumont (e.g., 1980, 1983). Dumont compares *homo hierarchicus* and *homo aequalis*, the former standing for what he terms the 'traditional' world and the latter for the West. For him, *homo hierarchicus* is encompassed in the collective totality, which underwrites his destiny; *homo aequalis*, on the other hand, is a self-contained, autonomous individual, taking his or her own decisions independently from society (Ingold 1986: 221; cf. Celtel 2005). Individualism or autonomy among hunter-gatherers, however, is of a different kind. In East Greenland, and similarly in other hunting and gathering societies, a person is, from the outset, embedded in relationships with others, and his or her autonomy, in terms of both intention and action, is conditional upon trust in support from others (see Ingold 2000: 69–70). Hence, 'the collectivity is present and active in the life of every individual' such that, as Ingold continues,

> there is no contradiction, no conflict of purpose, between the expression of individuality and his [a person's] generalised commitment to others. Since the world of others is enfolded within his own person, these are one and the same.
>
> (1986: 240, brackets added)

A relational autonomy has not only been found to be prevalent in hunting and gathering societies, but a number of theorists have also contended that autonomy is always fundamentally social in nature (see Christman 2004; Oshana 1998, 2006; Westlund 2008). Andrea Westlund emphasises that relational autonomy, 'does not require that one stands in idealized, egalitarian relations with others' (2008: 7). Describing personal autonomy as a matter of self-governance in choice and action, she writes that this self-governance 'point[s] outward beyond itself, to the position the agent occupies as one reflective, responsible self among many' (*ibid*: 7). A person is thus never 'independent' or without influence from other humans and, I would argue, from the non-human environment as well. On a general level, personal autonomy, as Marina Oshana asserts, is 'a matter not just of what goes on in an agent's head but also of "what goes on in the world around her"' (1998: 81). This relational autonomy is thus closely connected to a particular understanding of the person, upon which I elaborate below.

Personhood and selves

I understand a person not as a bounded, indivisible entity – characteristics that are often regarded as the hallmarks of the Western notion of the person – but as part of a dynamic process, constituted within dialogues with others, and being at the same time singular and multiple (Harré 1998: 20). Even the putatively 'Western' concept of a person is problematic, however, as it can exist only as an abstraction. On an ethnographic level, as Susan Rasmussen explains, the West 'is just as complex and internally differentiated as are those cultural settings outside of it traditionally designated as "non-western"', and there is no such thing as a monolithic western understanding of the person (2008: 36). Before evaluating other people's understandings of the person, we must thus problematise our own categories.

In the literature, one encounters diverse usages of the terms self, person, and individual, which are sometimes defined in different ways and at other times used interchangeably. I shall use 'person' to denote what others have called either self or person, arguing that in an anthropological enquiry the two cannot be distinguished on a conceptual level as their meanings differ from one cultural context to the other; distinguishing the self from the person right from the start carries the danger of projecting Western academic classifications and dualist paradigms onto other people (Morris 1994: 8; Rasmussen 2008).[3] Many early anthropological writings on personhood (e.g., in the work of Mauss, Hallowell, and Fortes) build upon a fundamental distinction between the self on the one hand, understood as the psychological unit and as the centre of self-awareness, independent of its involvement in relationships, and the person on the other hand, defined as the cultural conception – or 'category' (Mauss 1985) – of a particular community (Morris 1994: 10). Implicit in this distinction is the premise that social anthropology is the study of different concepts of the person formed within society (Carrithers et al. 1985; Morris 1994), and not of the self as the internally-defined locus of individual experience or the psychological substrate. According to this perspective, presupposing an opposition between self and society, the self is regarded as 'pre-social' (Ingold 1991: 356), a view that is in line with other fundamental dualities of Western thought. Yet many 'non-Western' people, Ingold asserts, tell us that there exists no self in advance of its entry into society (such that it becomes a person only thereafter), but rather that the person, just as the organism, is – right from the start – embedded within social-relational contexts (*ibid*: 367, see also Myers 1986; Rosaldo 1980). This corresponds to the East Greenlandic understanding of a person, as I will show in this book. My field material suggests that a person in East Greenland constantly grows within the process of social life, a process which does not start at birth in order to end at death, but is connected to a far wider relational field. Hence, I follow Ingold's conclusion that

> personhood is no more inscribed upon the self than it is upon the organism; rather, the person is the self, not however in the western sense of the private, closed-in subject confronting the external, public world of society and its relationships, but in the sense of its positioning as a focus of agency and experience within a social relational field.
>
> (*ibid*: 367)

Though nowadays only a few social scientists would still argue that the self is a 'pre-social' entity, and most studies speak of a relational self, the categorical distinction between person and self still underpins much contemporary writing. For instance, Rom Harré has proposed to distinguish 'between the individuality of a human being as it is publicly identified and collectively defined [i.e. person] and the individuality of the unitary subject of experience [i.e. self]' (1984: 76, brackets added). He thus justifies the division between self and person with reference to the public/private distinction, not however taking into account that the distinction between private and public realms is also culturally relative (Rosaldo 1974). Yet, in one of his later works, Harré refines his position, arguing that '[t] here are only persons. Selves are grammatical fictions, necessary characteristics of person-oriented discourses' (1998: 3–4). Selves, accordingly, are 'aspects of persons' (*ibid*: 5). Brian Morris concurs with this view by declaring that the self is an abstraction, produced by a human person, and that it refers to a process rather than to an entity (1994: 12). Much in line with the latter view, I use 'person' as the fundamental notion, arguing that depending on the cultural context, it may or may not be set in contrast to some kind of self.

Closely related to these discussions, a number of authors such as Marilyn Strathern (1990), Tim Ingold (1999b), and Michelle Z. Rosaldo (1980, 1984) have questioned the assumption that at the heart of every society lies the antinomy between society and individual. In her work on Melanesian societies, Strathern introduced the notion of the 'dividual' person as a multiple partible agent who 'contain[s] a generalized sociality within', and which she sets in contrast to the Western idea of the indivisible and unitary person (1990: 13). The Melanesian person, she asserts, has to be imagined as a 'social microcosm' and 'a derivative of multiple identities' (*ibid*: 15). Some authors have (rightly) observed that under the guise of the distinction 'Western' versus 'Melanesian', which are indeed idealised categories, Strathern is comparing an abstract category with an ethnographic reality (Hess 2006: 286). Nevertheless, anthropologists and other scholars who find similarities among different peoples around the world have picked up her concept of the 'dividual' person (e.g., Bird-David 1999). On a broader level, it bears comparison with a number of similar approaches that highlight the relationality of a person, for example discussions about the 'dialogical self' (leading back to Mead 1934), or Lave's (1988) notion of the 'person acting'.

Slightly broader than Strathern and other scholars who mainly refer to *social* contexts when speaking about relational personhood, and much in line with my own approach, Jean Lave emphasised the 'fundamental priority of relatedness among person and setting and activity' (1988: 180). She argues that various aspects of the 'lived-in-world' constitute the embodied person, yet that the person's direct involvement with his or her environment has often been neglected. This argument parallels other criticisms of social constructivism, a line of thought that has for long prevailed within anthropological circles and still frequently appears therein. Contrary to conventional theories (including many social constructivist approaches) that have often conceived of the person's setting and activity only in terms of their location within the field of a person's (cognitive) representations, Lave argues for a theory of practice in which 'setting and activity connect with mind through their constitutive relations with the person-acting' (*ibid*: 180–1). This closely parallels Ingold's ecological approach and his notion of 'direct perception' (e.g. 1993a, 2000), which he draws from the work of J.J. Gibson (1979).

A further important point, highlighted by both Lave and Strathern, is that personhood is not necessarily bound to the physical body, and that, as Lave writes, 'The person-acting and social world as mutually constituted are not always or exactly divided by the surface of the body' (1988: 181). The person may have permeable boundaries, and it may be a 'partible' entity 'that can dispose of parts in relation to others', as Strathern shows in her work on exchange in Melanesia (1990: 185). She demonstrates how items may circulate as parts of a person (*ibid*, cf. Gosden and Knowles 2001), and Christopher Tilley (2004) argues that places and landscapes may also become parts of persons. These studies epitomise a general trend in social science research which points to the ambiguities in the boundaries of corporeality itself, questioning the view of the 'physical' body as a bounded entity. Thomas Csordas has argued:

> Our lives are not always lived in objectified bodies, for our bodies are not originally objects to us. They are instead the ground of perceptual processes that end in objectification.
>
> (1994: 7; cf. Merleau-Ponty 1962)

These studies reveal that knowledge is always grounded in a particular embodied and situated standpoint (Haraway 1988), and that culture and experiences have to be understood from the standpoint of bodily being-in-the-world (Csordas 1999: 143; cf. Jackson 1996).

Moreover, and as mentioned above, from the perspective of particular groups of people, personhood may also encompass non-human beings and entities (e.g., Harvey 2005). According to 'animistic' understandings, particular animals or environmental features may be alive, have intentions, and be considered as persons (see for example, Hallowell 1955, 1960, on the Ojibwa; or Bird-David 1999 on the Nayaka), a view which also appears in East Greenlanders' lifeworlds.

In order to illustrate my use of the term 'person', and the theoretical discussions it builds upon, I have offered only a brief glimpse of theories of self and personhood, concentrating on approaches that are particularly relevant for East Greenlandic understandings of the person. I want to stress that in this book I will not focus on cultural representations or cognitive concepts of personhood, which I regard as inseparable from social praxis, or on discourses about personhood, but on the ways personhood appears in people's practices, and, in particular, in people's communicational practices.

Communication: a creative process and its multiple modalities

This book explores interpersonal communication among East Greenlanders in the Ammassalik region and the varied channels of nonverbal, material as well as linguistic interaction. Following Ruth Finnegan, my view of communication is 'not confined to linguistic or cognitive messages but also includes experiences, emotion and the *un*spoken' (2002: 5). Communication is regarded as a creative process rather than the transport of data or 'meeting of minds', and goes beyond the preoccupation with information, transmitter, and receiver that – firmly anchored in Saussurean structuralism – has shaped much of the current discussion (e.g., Shannon and Weaver 1964). It entails the manifold modes of persons interacting and living, both near and distant – through smells, sounds, touches, sights, movements, embodied engagements, and material objects (Finnegan 2002: 5; and see Birdwhistell's [1970] view of communication as 'multi-channel'). I argue that communication is a relational phenomenon which, apart from the communicating parties and other human actors, involves the broader contexts and settings of the communicative encounter. As Ingold has explained, 'there is no "reading" of words or gestures that is not part of the [person's] practical engagement with his or her environment' (1997: 249, bracket added).

In order to explore communicative encounters among the Tunumeeq, I draw from studies of both sociolinguistics and linguistic anthropology, which predominantly consider language when speaking about communication (e.g. Duranti 2009), and approaches dealing with the body and the senses. On the one hand, as regards the former, contributions from the *ethnography of speaking* pioneered by Dell Hymes (1962, 1964), nowadays mostly referred to as the *ethnography of communication* (Gumperz and Hymes 1986; Hymes and Gumperz 1964; Saville-Troike 2003; Sherzer 1983, 1990), have been particularly relevant. This body of work, which interrelates the study of language and culture, asks what the speaker needs to know in order to communicate appropriately within a particular speech community, thereby considering both communicative competence and performance. In relation to this, it has repeatedly been emphasised that language cannot be separated from how and why it is used (Saville-Troike 2003: 1–2). The performance-centred approach also highlights the important role of the audience in the communicative encounter, a point which has been at the centre of a number of schools of thought, ranging from Bakhtin's dialogism (1981), to Goffman's social interactionism (e.g., 1959, 1967), to Watzlawick's pragmatics of human communication (Watzlawick et al. 1967), to conversation analysis (Goodwin 1990).

On the other hand, considering various communicational modalities apart from speech, I draw from studies of so-called nonverbal communication, including Hall's comparative approach on proxemics and use of space (1959, 2003), and Birdwhistell's (1970) work in kinaestethics and body motion. Stressing the multisensory aspects of communication, I further rely on contributions from the anthropology of the senses, which only partly coincide with studies of so-called nonverbal communication mentioned above. Amongst others, I draw from Cristina Grasseni's work on skilled vision (2004, 2007a, 2007b), and Farnell's studies of body movement (e.g., 1995, 2003).

Speaking about the senses, I am aware that the distinction between five primary senses, with an additional kinaesthetic sense, has its roots in the Western philosophical tradition, and that sensory perceptions and ways of experiencing in different cultural contexts are not necessarily based on such differentiation (cf. Classen 1993: 50–76; Farnell 2003; Ingold 2000: ch. 14). Even in a Western context, as Constance Classen shows in her investigation of our everyday language, we find a certain interchangeability of sensory meaning, which suggests 'that sensory perception was once conceived of in a more fluid fashion than nowadays, with less rigid distinctions between senses' (1993: 56). Likewise, different communicational modalities such as speech, gestures, facial expressions, and so forth are often inextricably interlinked and cannot be explored apart from one another.

This book, thus, does not aim to single out speech as the primary mode of communication, followed by a range of subordinated, 'bodily' communicational modalities, nor does it merely focus on either the former or the latter. Exploring the relevant literature, however, one encounters few contributions that combine a focus on language and speech with the study of so-called nonverbal communication (but see Finnegan 2002). This separation of the verbal from the nonverbal rests on a division between mind and body, rendering language as a cognitive and symbolic endeavour, or, at least, categorically different from other ways of communicating (Farnell 1999: x), a distinction which downplays the diversity of communication other than speech, and overstates the extraordinary position of language (see Farnell 1995, 1999; Ingold 1997; Sheets-Johnstone 1999, 2009). Contrary to some cognitive approaches that consider words mainly in terms of messages that carry meaning in themselves, language itself has to be regarded as an embodied practice which involves the constantly moving body. As Farnell has explained, 'social actors consistently and systematically use bodily movement as a cultural resource in discursive practices and not simply in addition to them' (1999: viiii). On a broader level, thus, 'Far from carrying meanings *into* contexts of interaction, words – like gestures – gather their meanings from the contexts of activities and relationships in which they are in play' (Ingold 1997: 249; see Merleau-Ponty [1964] on the relations between vision and movement).

In addition, conventional models of communication tend to assume that it takes place in face-to-face encounters, such that 'information' passes back and forth between static communicators. Yet, if we regard movement as integral to all kinds of communication, be it with respect to people literally changing place, moving the mouth, head, or drawing a gesture,[4] then the communicational field broadens to encompass different ways by which people recognise each other, and positions through which they do so. This may entail communication through directly looking at each other, via sideways glances (Downey 2007), or by grasping a particular body part – or something perceived as an extension of the person – in some way. Communication may then be defined as 'mutually recognized actions' (Finnegan 2002: 3), with the actual feedback between the communicants distinguishing communication from mere perception.

My rather broad view of communication builds, amongst other sources, upon Ingold's work on perception and his 'anthropology of the line', which rests on the premise that we do not perceive from a fixed point of view but along a 'path of observation' (2004a, 2007, 2011). Ingold reasons that if perception is a function of movement, then what we perceive – and I would add, what we communicate – must also depend on how we move. Our movements as we go along are 'continually and fluently responsive to an ongoing perceptual monitoring' (Ingold 2004a: 332). This 'tuning' also happens with regard to other persons: eye-to-eye-contact, for example, is an experience of attunement (Sheets-Johnstone 2000: 348). And whilst walking together people adjust their movements to the movements of their companions, with communication involving the whole body (Lee and Ingold 2006: 80). Following Maxine Sheets-Johnstone (2000), this 'joint attention' has to be regarded as a basic human phenomenon and attests to a general 'intercorporeal awareness'. Communication through joint attention and a transfer of sense involves 'grasping the integral body of the other intuitively' (*ibid*: 348).

My notion of communication as a 'joint attunement'[5] will not be confined to humans only, but may also, for example, be applicable to animals (cf. Bateson 1972: 364–78). Studies of animal communication have shown that, 'communicative behaviour is not a human monopoly' (Griffin 2001: xi), and that, as Donald Griffin argues, many animals may express feelings and simple thoughts, which are reflected in the ways they communicate. Although whether or not they are able intentionally and consciously to communicate is much debated (*ibid*: xi, Ingold 1994: 6), in this anthropological study I am less concerned with scientific questions of consciousness and the like, and more with East Greenlanders' understandings of them. Like Griffin, Ingold asserts that language is not in the first place an instrument of cognition, thereby supporting a broad view of communication, which is not limited to humans.[6] I follow his argument that language is not on a higher evolutionary level than systems of nonverbal signs, and that it may therefore be regarded as one 'species-specific mechanism of communication' among many (Ingold 1994: 7).

Indirect communication

Everyday communication in East Greenland encompasses manifold implicit and subtle modalities in ways of speaking, gestures, facial expressions, bodily postures, and different types of meaningful behaviour, which in the literature have often been subsumed under the label of 'indirect communication' (Hendry and Watson 2001). Though I see some usefulness in the underlying concept, it is important to point out that the line between the categories of 'indirect' and 'direct' communication is itself culturally specific, and might sometimes not be drawn at all. Hence, this distinction does not always prove useful when looking at different peoples' communicational practices, and what ethnographers might sometimes call 'indirect' could, among the people in question, possibly be understood very clearly and directly, particularly when it comes to various nonverbal modes of communication. In the East Greenlandic context, I came across many examples that fit the category of indirect (or implicit) communication well. But there were many other instances where I found it difficult to separate direct from indirect communication, and where any such distinction would have been inappropriate.

Due to the relevance of this distinction for my study, nevertheless, I want to elaborate briefly upon some important contributions to the literature dealing with indirect communication.

Though the concept is much in evidence in both academic and non-academic literature, its definition has often been taken for granted or formulated rather vaguely. Søren Kierkegaard (1992) was one of the first to provide a more detailed description of the notion of indirect communication (Kellenberger 1984). For him, indirect communication reduplicates a subjective truth and is characterised by a particular inwardness, in contrast to direct communication which he assigns to the field of 'objective thinking'. The latter presupposes certainty, whereas the former takes place in the 'process of becoming' which does not (yet) allow for certainty (Kellenberger 1984: 153–4). Kierkegaard (1992) argues that, with regard to indirect communication, it is the action that is important, along with the fact that the interpretation of the communicative act lies with the addressee and not the addresser. Not wanting to deny the usefulness of Kierkegaard's work, I am critical of some of his premises, above all his juxtaposition of subjectivity and objectivity, which arises from the premise that there exists something like a truly 'objective' truth, independent of a particular standpoint. Moreover, with respect to the certainty he assigns to the field of direct communication, one could ask where one should allocate those cases of communication where someone expresses a particular uncertainty without disguise or indirection of any kind.

Donald Brenneis (1987) has defined indirection in a slightly different manner than Kierkegaard. He explains that, first, indirect communication is based upon the fact that 'speakers "mean" more or other than what they say', with the message going beyond the literal – a point which appears in most other definitions (Brenneis 1987: 504). Second, he argues that 'indirection usually allows the avoidance of responsibility', which is closely connected to his third point, in which he stresses that addressees 'are not only allowed but compelled to draw their own conclusions' (*ibid*: 504). This point was also mentioned by Kierkegaard (see above). Finally, Brenneis asserts that there is a formal element involved in indirection, which signals that 'more is going on than meets the ear' (*ibid*: 505).

Many examples in the literature have shown that it is possible to be indirect in various ways and to varying degrees, be it through language and particular ways of speaking, through gestures, facial expressions and bodily postures, through various types of behaviour, or through the help of intermediaries of diverse sorts (see, for example, Hendry and Watson 2001). Indirect communication figures to some extent in all societies, though the modes of communication, as well as the relationship between the said and the unsaid, can differ substantially (Brenneis 1987: 503; Hendry 1989: 623). As Robin T. Lakoff writes, 'fully competent communicators have access to both [direct and indirect communication], actively and passively, and know where and why to use each according to the rules of their culture' (2007: 130, brackets added). Lakoff further stresses that one of '*our* culture's basic beliefs' – presumably either referring to her American background or to the problematic conundrum of Western culture – is that one should be honest, make oneself clear, and communicate directly. This widespread ideal, she argues, is restricted by situations of trying not to hurt others or to infringe on manners and taste (*ibid*: 131, my emphasis).

> So the absolute certainty of being understood, while a conversational desideratum, is overshadowed for most people in most kinds of discourse by the desire for protection, both self-defence and solicitude for the self-esteem of others.
>
> (*ibid*: 132)

The ideal of communicating directly, however, is not universally applicable, and in East Greenland, for example, one encounters diverging forms of politeness and manners of conduct, some of which could be called direct, or indirect, and others which are not clearly allocable to one category or the other. The desire for protection to which Lakoff draws attention, however, has been noted in many ethnographic accounts, and may be a way of avoiding an intrusive communicative act, such as an order, by replacing it with a less intrusive communicative mode, such as a question or a facial expression (*ibid*: 130). In this regard, however, one has to take into account that people's perceptions of what is an intrusion may differ cross-culturally, which may be linked, amongst other things, to the value of personal autonomy and varying notions of personhood, as I will show in the case of East Greenland.

Many authors have contended that indirection has a face-saving function, for either the addresser or addressee (e.g., Hendry 1989; Schottman 1993). Wendy Schottman explains:

> If it is the addressee's face that would be threatened – by criticism, lack of respect, or an impingement on freedom or privacy – then the act must be disguised so as not to risk provoking an open conflict. If, on the contrary, it is the speaker's face that would be threatened – by a request that could be refused or by the speaker manifesting a lack of restraint and modesty while speaking – then the act must be disguised in order to protect the speaker's reputation.
>
> (1993: 541)

In this, Schottman builds on Brown and Levinson's (1978, 1987) argument that there are universal features connected to face and politeness. Brown and Levinson define face as the public self-image that every member of a society wants to claim for him- or herself, drawing here from Erving Goffman (1955, 1967), who defines face as the positive self-image, in the sense of a kind of mask, the delineation of which depends on the audiences and type of social interaction. Face has to be protected in order to maintain dignity and honour. Brown and Levinson, moreover, differentiate between negative and positive face, understood as 'the desire to be unimpeded in one's actions (negative face), and the desire (in some respects) to be approved of (positive face)' (1987: 13). Brown and Levinson regard the concept of face as universal, though subject to cultural variability, as Igor Strecker (1993) shows. I do see some value in thinking about a person's 'face' in terms of personal integrity, dignity, and honour. All in all, however, I am critical of some of the underlying premises of this concept, in particular concerning the presupposition of an 'inner' self set against a social person, and the distinction between private and public upon which the concept builds. The idea that the face is a mask hiding the 'real' person within is founded on a certain concept of personhood (bounded and self-contained vis-à-vis an external public arena), which I have criticised above.

Let me return to the theme of indirect communication. Schottman (1993) pro-poses four subdivisions of indirection, though these are mainly limited to verbal indirection. She distinguishes between: 1) indirectly formulated verbal communi-cation; 2) indirectly addressed verbal communication, e.g. 'The real addressee is one of the overhearers' (*ibid*: 542; 3) verbal communication that has an indirect author; 4) verbal communication that is indirect because of its 'key', such as, for example, jokes. Another categorization has been proposed by Donald Brenneis, who differentiates between 'text-centred' indirection, referring to features of the message itself, 'voice-centred' indirection, as regards the ambiguity of who really is the speaker, 'audience-centred' indirection, which deals with the question of who in the audience is intended to be influenced, as well as indirection 'based in events', with the event's occurrence itself being particularly meaningful – thereby including indirection through communicative acts other than speech (1987: 505–7).

Indirection has been found in different kinds of societies, both hierarchical and egalitarian in nature. For instance, James Scott (1985, 1990) has shown that indirection may provide a means for peasants, or subaltern people more broadly, to articulate their resistance against colonial powers, and Schottman (1993) has illustrated how indirection among the Baatombu in Benin is used to reproach both superiors and people of equal status. Having worked among the Baatombu as well, Erdmute Alber (2004) relates the avoidance of confrontation and forms of indirect communication to a lack of institutionalised ways of dealing with con-flicts (referring here to the work of Georg Elwert). In addition, many ethnographic accounts stress that indirect communication is particularly pronounced in societ-ies where egalitarian notions strongly inform social life, which is also supported by my fieldwork in East Greenland (see Brenneis 1987; Morris 1981; Williams 1987). Brenneis affirms that 'a concern for one's own reputation and freedom of action and a preoccupation with those of others', which often goes along with a strongly egalitarian ethos, nurtures indirection (1987: 508). In another article, he argues that the dilemma of both acting politically and not showing that one does so is particularly marked in societies 'in which clear-cut leadership does not exist and decision making is consensual' (Brenneis 1984: 70), as is the case in East Greenland. Indirect communication, here, is associated with the attempt to avoid 'the perils of direct leadership and confrontation' (*ibid*: 70). Communicational modalities subsumed under the label 'indirection' may thus speak directly to the conditions of life in particular communities (Alber 2004), which should, however, not be taken as an overall explanatory framework.

Taking into account the relationality of East Greenlanders' communicative practices – some of which might be approached by thinking in terms of direct versus indirect communication – I will explore the embeddedness of everyday communication in East Greenland within various social, cultural, and environmental spheres of daily life. This relational approach opens up a variety of issues and themes, including leadership and decision-making, kinship, human-environmental relations, and so forth. The particular prominence of indirect forms of communication in East Greenland, I will argue, is closely connected to the high value placed on personal autonomy, which again relates to a particular understanding of the person. This argument will be developed by – now and then – pointing to similarities in interpersonal communication with other small-scale societies, many of which are subsumed under the label of hunter-gatherers or that are (partly) egalitarian in nature, whilst at the same time trying to convey the specific East Greenlandic features of communication and inhabitants' understandings of personal autonomy and personhood respectively.

Notes

1 Throughout the last decades there has been much discussion of the usefulness of the category 'hunter-gatherers', mostly relating to the fact that nowadays hardly any group of people still exclusively lives from hunting and gathering (Myers 1988). However, though hunting and gathering have often been supplemented with, or replaced by, other economic strategies, many of these people show commonalities in their particular relationship with their environment and fellow humans, based on sharing, autonomy, and immediacy (Bird-David 1990; Ingold 1999b).

2 When speaking of the 'West', I refer to an ideological category, devoid of any geographical connotation, which entails the technocratic belief in rational inquiry, underpinned by various dualities such as mind/matter and nature/culture. Though these dualities have indeed been shown to be highly problematic (e.g. Latour 1993), even in application to nominally 'Western' societies, the term is sometimes difficult to avoid, especially since the academic endeavour of this book also builds on some of these 'Western' ideological assumptions (Ingold 2000: 6).

3 I use 'individual' in a rather broad and all-encompassing way, particularly when distinguishing between groups of people and individual actors, without implying any specific theoretical approach.

4 Cf. Farnell (1995: 6), who speaks of 'vocal gestures' and 'manual gestures'.

5 My use of the terms 'attunement' or 'tuning' should not be taken to imply that communicants necessarily agree with each other; communication in terms of a 'joint attunement' may entail both agreements and disagreements between the communicating parties.

6 Ingold, however, calls into question Griffin's argument that animals first think (or come up with a plan or design) and then act or execute something. He argues that non-human animals do not think before they act when even human beings so rarely do so, and that thinking for both humans and non-humans is rather *in* the action (Ingold, personal communication, 25.10.2010, cf. Suchman 1987, on situated action).

2   East Greenland

Historical and ethnographic background

The Ammassalik region in East Greenland is cross-cut by the Arctic Circle which lies approximately 100 km north of Tasiilaq (see Map 2.1). Like the whole Greenlandic east coast, the region is distinctive for its cold sea current running down from north to south, which brings along huge amounts of ice from North Greenland and the North Pole region. This makes navigation for long periods of the year extremely hazardous and often impossible. Only during approximately four months of the year can the region be accessed by boat; provision ships from Denmark deliver groceries and various goods from July to mid-November (depending on ice conditions). The rest of the year the region is accessible from West Greenland and Iceland via the airport in Kulusuk and connecting helicopters to Tasiilaq and the other villages.

The Greenlandic east coast has approximately 3,300 inhabitants. This is a low population density compared to West Greenland, and in relation to Greenland's overall population of about 56,000 people. For the two districts of East Greenland, Statistics Greenland (2017) records 2,911 inhabitants for Tasiilaq and 375 for Ittoqqortoormiit. Due to their isolation because of the pack-ice, the population of the east coast did not have sustained contact with Southerners for a long time. Contrary to West Greenland, where colonisation began in 1721, colonisation in the Tasiilaq region started as recently as 1894.

Historical developments of the region

As stated in most written sources about East Greenland, sustained contacts between the inhabitants of the Ammassalik region and non-Greenlandic people started with Gustav Holm's arrival in the area in 1884.[1] The Danish naval officer and his crew were searching for remnants of a former Viking settlement (from medieval times), which from its name 'eastern settlement' was suspected to have been located along the east coast of Greenland. (Later it was found out that this settlement was actually situated at the very southern tip of Greenland.) Holm and his crew, unlike other expeditions at that time, were using *umiat* (sg. *umiaq*), Greenlandic women's boats, for their journey. These large boats, covered with seal skins, are well adapted to the difficult ice conditions. Thus, the Holm expedition became the first group of Europeans who managed to reach the Ammassalik area. Instead of

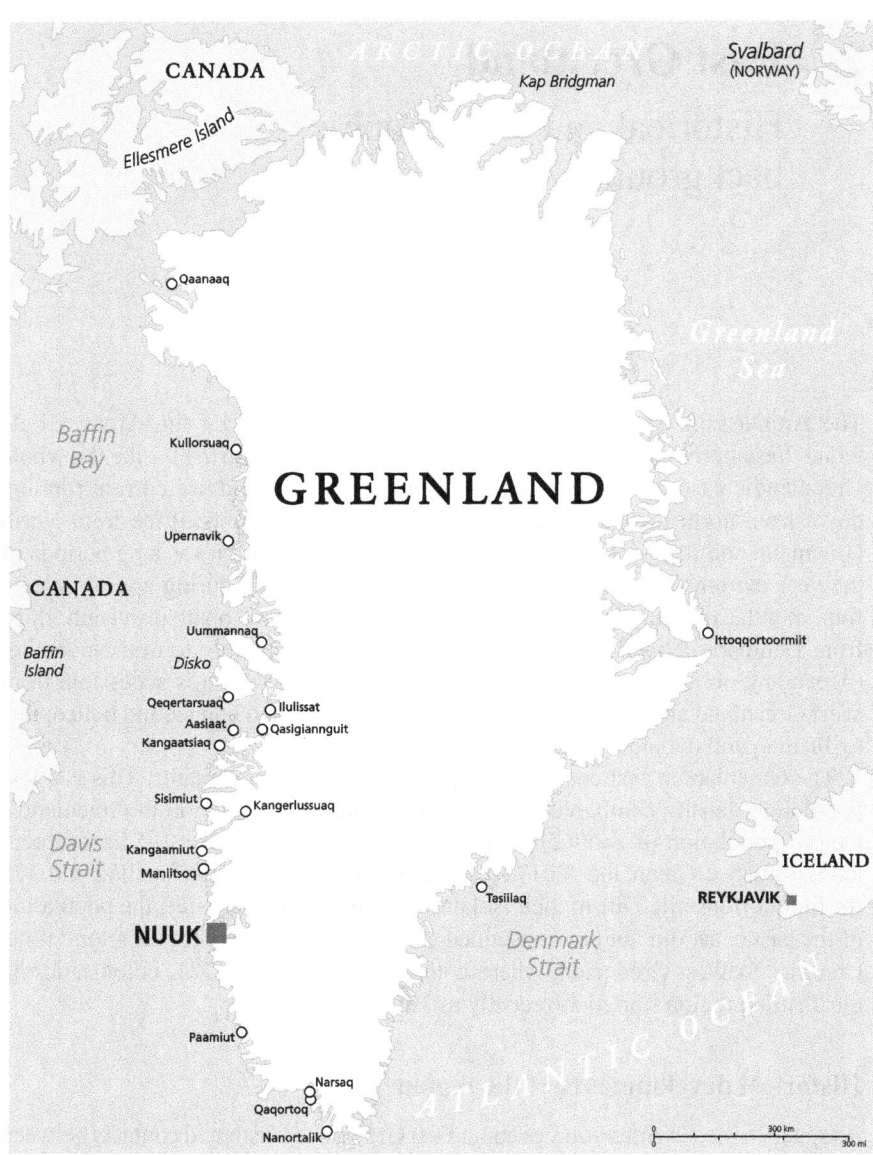

Map 2.1 Map of Greenland
(© PeterHermesFurian/iStock)

discovering traces of the Vikings, they found a semi-nomadic Inuit population, scattered along the fjords of the area, which numbered 413 individuals at that time. The crew stayed for ten months. Thereafter, Holm wrote the first ethnography of these people, who called themselves Iivit, giving a detailed description of a wide range of aspects of society, such as material culture, social life, and so-called intellectual culture (Holm 1888, 1911).

Ten years later, Holm returned to the area to set up a trading station and missionary post. He was accompanied by the Johan Petersen, who was responsible for administration and trade, and the missionary Frederic C.P. Rüttel. The administration of political, social, and economic issues was now carried out directly from Copenhagen, with local affairs being managed by the colonial administration in Ammassalik (nowadays the town is called Tasiilaq). By 1921, the whole population was baptised Lutheran. Settlements became more permanent, and the colony experienced an increase in population numbers. In 1925, the colonial administrator Ejnar Mikkelsen thus initiated and effected the transport of seventy Iivit some 1000 km further north (followed by more people in the subsequent years) to establish the settlement of Ittoqqortoormiit (Scoresbysund) (Mikkelsen 1925, 1944). The establishment of this new Greenlandic settlement was aimed at both alleviating the population problems in the Ammassalik area and affirmed Danish claims in the conflict with Norway over the sovereignty of East Greenland (Robert-Lamblin 1986).

The rapid economic and social-cultural developments that have characterised recent Ammassalik history did not commence until the Second World War. Before then, major cultural changes had been avoided; the population had been largely shielded from much contact with the outside world, and the colonial administration had supported a local seal hunting economy. The Danish authorities wanted to protect the small ethnic group, partly from diseases for which they lacked immunity, and partly from the risks related to abandoning their 'traditional' lifestyle. With the Second World War, however, Denmark's protection policy and isolationism could no longer be sustained. East Greenland depended on the United States for a period of five years, and the U.S. government set up a radio-meteorological station in Tasiilaq in 1941. In 1942, they built the military base in Ikkatteq, in order to refuel airplanes between America and Europe. The flux of American soldiers and workmen brought about cultural changes of various sorts, including the introduction and use of consumer goods (Robert-Lamblin 1986: 12).

Yet by 1950, as Robert Petersen asserts, though '[s]hamans and other distinguished individuals had disappeared, . . . the society, cemented not by leaders but by the reciprocity of free hunters, was still functioning' (1984b: 639). Major cultural and economic changes in the region took place between the 1950s and early 1960s. The population was regrouped into larger villages and and wooden houses gradually replaced the turf and stone houses that the East Greenlanders had lived in before. The isolation of the Ammassalik area ended when the airport in Kulusuk was built in 1957, and construction work brought in a considerable number of Americans and Danes for several years (Elixhauser in press-a). Commercial aviation commenced in the early 1960s (Robert-Lamblin 1986: 91).

The modernization of the small capital of Tasiilaq began at around the same time, connected to an influx of Danish residents. Some inhabitants took up wage earning positions, in the fishing industry (with commercial cod fishing in Kuummiut established in the late 1950s), or for the local administration. 'Westernization' efforts continued throughout the 1960s, linked to educational policies that aimed to integrate Greenlandic youth into Danish society (Nooter 1972/73, 1984a, 1988; Petersen 1984a).

In 1963 a municipality system, like that in West Greenland, was established in East Greenland, and throughout the 1970s increasing autonomy was granted to various political bodies. This went along with diverse efforts to achieve greater independence from Denmark. Home Rule (*hjemmestyre*) was celebrated in 1979, with Greenland taking over its own administrative affairs. Self Rule (*selvstyre*) was granted in 2009.

Hence, in East Greenland transformations from a hunting society to a society supported also by wage labour have taken place within only a few decades, and at a much greater speed than developments in West Greenland, where colonisation started as early as the 1720s. Though East Greenlanders have quickly adapted to some of these developments, especially in material culture and technology, and through a number of customs and traditions have been adapted to the new circumstances, these rapid changes have also had unforeseen consequences, which become apparent when looking at the comparatively high rates of alcohol abuse and violence (e.g. Robert-Lamblin 1984). Today, one still encounters substantial economic and social differences between East and West Greenland.[2]

Ethnographic research

East Greenland has attracted a comparatively high number of ethnographers since it was first colonised. The early sources, based on research conducted in the early decades of the 20th century, were based on the then common endeavour of trying to describe a culture in its totality, by exploring a variety of themes, ranging from social life, material culture, and economic activities, to oral history, just to name a few. A main research objective was to collect and preserve the traditions of a culture seen to be in danger of 'dying out'. These accounts, first and foremost, include Holm's renowned *Ethnological sketch of the Angmagsalik Eskimo* (1888, 1911) as well as William Thalbitzer's ethnographic writings (1914b, 1921, 1941), which are important ethnographic sources still today. Thalbitzer also conducted the first extensive research on the East Greenlandic language (1914a). Moreover, anthropologist Paul-Émile Victor left substantial ethnographic material based on his fieldwork during the 1930s (e.g., Victor et al. 1991; Victor and Robert-Lamblin 1989, 1993), and Knud Rasmussen collected legends and myths (1921; see also 1938). From around the middle of the 20th century onward, ethnographers gradually limited their enquiry to more particular socio-cultural issues, instead of striving for the impossible goal of giving an all-encompassing account of 'a culture'. In the following, I will briefly sketch some major themes in the ethnographic literature on East Greenland.

Research into oral history, started by Rasmussen, Holm, and Thalbitzer, continued throughout the 20th century, with contributions, amongst others, from artist Jens Rosing (1960, 1963, 1993) and anthropologist Birgitte Sonne (1986, 1990). The latter has also written on a wide range of social issues from a historical perspective (Sonne 1982, 1986). A prominent theme in the ethnographic literature is cultural changes connected to the rapid societal developments, discussed by various authors such as anthropologists Gert Nooter (1972/73, 1984a, 1984b, 1984c, 1988) and Cunera Buijs (1993; see also Buijs and Oosten 1997), Robert Petersen (1984a, 1984b, 2003), and the anthropologists Robert Gessain (1967, 1969) and Joëlle Robert-Lamblin (1986). Robert-Lamblin has also written on kinship and family life, gender as well as various social issues (1981, 1984, 1992, 1997a, 1999a, 1999b, 2000). Some of these themes have also been dealt with by other female researchers, such as Cunera Buijs (2002) and Bernadette Robbe (1975a), the latter conducted long-term ethnographic fieldwork together with her husband Pierre Robbe. Pierre Robbe focused his research on environmental knowledge and human-environmental relations, elaborating upon subsistence activities such as hunting and various themes related to it (Robbe 1977, 1994). The broader theme of human-environmental relations has also been explored by the anthropologist Grete Hovelsrud-Broda, who wrote about the integrative role of seals in the village of Isertoq, and who has examined the effects of the anti-sealing controversies in the 1980s on the local economy and kinship structure of the village of Isertoq (Hovelsrud-Broda 1997, 1999a, 1999b, 2000b).

Another prominent topic in the ethnographic literature on the region is material culture and concomitant changes. This body of work includes first and foremost the aforementioned work of Gert Nooter, who has written on a wide range of topics such as leadership (Nooter 1976) and cultural change (see above), mainly focusing on the relationship between people and material objects (based on the assumption that 'when things are changed people change too' [Nooter 1984a: 143]). In more recent times, this research tradition, giving materials a central role in the study of culture, has been followed up by another Dutch anthropologist, Cunera Buijs, who conducted research on clothing and identity in Tiniteqilaq (2004; Buijs and Petersen 2004) and more recently on climate change (Buijs 2010). French researchers have also investigated the use and meanings of material objects, for example Robert Gessain, who, in some of his writings, collaborated with Paul-Émile Victor (see Gessain 1968, 1984; Gessain and Victor 1969, 1973, 1974). In addition, anthropologists have written extensively on topics such as shamanism and sorcery (e.g., Robbe 1983; Robert-Lamblin 1996, 1997b),[3] religious conversion (Hindsberger 1999), myths and cosmological conceptions, and a number of related issues, such as naming practices (e.g., Gessain 1978, 1980; Robbe 1981). After this short overview of the literature, I will give some background information of the main locations of my research, the regional capital Tasiilaq and in greater detail the village Sermiligaaq (see Map 2.2).

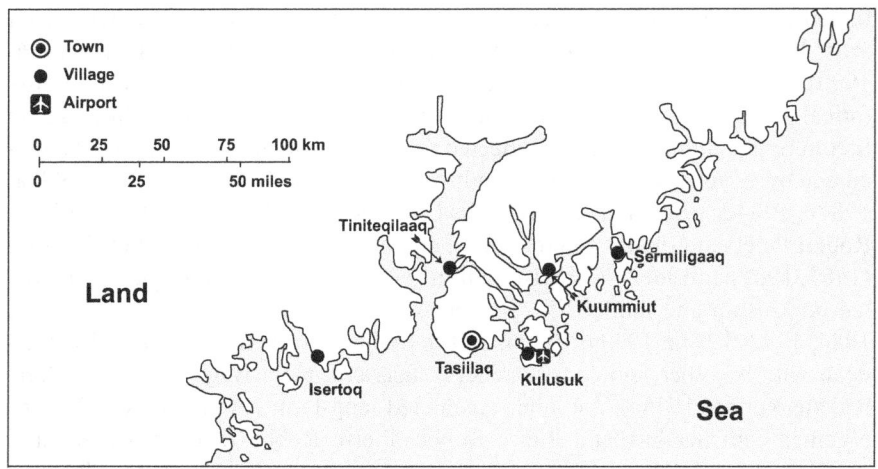

Map 2.2 Map of the Ammassalik region, East Greenland
(Illustration: Martl Jung)

Tasiilaq

The town of Tasiilaq, which was officially called Ammassalik for several years, is the biggest settlement on the Greenlandic east coast (see Figures 2.1 and 2.2). In 2017, it had a population of 2010 people (Statistics Greenland 2017). Until 31 December 2008, Tasiilaq was the administrative centre of the Ammassalik district. Thereafter, Ammassalik merged with four other former municipalities into the new Sermersooq municipality, with its capital in Nuuk.

The town offers a range of job opportunities, in public administration, at Greenland Trade (KNI – Kalaallit Niuerfiat), operating the two *Pilersuisoq* shops in town, tourism business (especially during summer), the social and educational institutions, the fishing industry as well as cargo and transport businesses, and a number of private companies, including the two hotels. Many of these jobs are unskilled, and the turnover of people between different posts is high. Skilled posts are often taken by a relatively small number of educated Greenlanders and Danes[4] on account of the low level of education of many East Greenlanders. Unemployment rates are high.

Tasiilaq accommodates a Danish community of about 5 to 10 percent of the total population; Danish residents tend to work in the schools, the hospital, in administrative positions for the municipality, or in one of the private companies in Tasiilaq. Apart from a limited number of long-term Danish residents (or Greenlanders with a Danish background), Danes often stay for only one or two years, after which they return to their home country. In addition, one usually encounters a number of short-term Danish workmen (sometimes including other nationalities), such as construction workers, employed for specific projects lasting for only a few months.

Figure 2.1 The old part of town, Tasiilaq, April 2005
(Photo: Sophie Elixhauser)

Figure 2.2 Tasiilaq school and blocks of flats, July 2007
(Photo: Sophie Elixhauser)

Tasiilaq serves as a hub for people from the surrounding villages, due to its economic opportunities, as well as its educational and social institutions. Moreover, pupils from the villages move to Tasiilaq at the age of 14 for their last two years of school education (*collegia*) (see Figure 2.2). Possibilities for continuing with formal education on the Greenlandic east coast are scarce (the few possibilities include a training in health care), and for most types of training or schooling East Greenlanders have to move to the west coast or to Denmark.

Tasiilaq has a road system, which is limited to the town area and does not connect to other places. It is extensively used by cars (and some taxis) during summer time, and by snow-machines in the winter. Transportation outside of the town mainly takes place via helicopter (the airport is situated in Kulusuk), or boat in the summer. Tasiilaq heliport offers helicopter services to the airport in the village of Kulusuk, and to the four other regional villages. In Kulusuk, airplanes connect to the Greenlandic west coast (Kangerlussuaq, Nuuk), Ittoqqortoormiit in Northeast Greenland, and Iceland. Helicopters operating between Tasiilaq and the regional villages approach each village once a week in the summer, and twice a week in the winter. During ice-free times, most Tunumeeq prefer to travel by private motorboat, if available, which is much cheaper. From July to November, another economic travel option is a boat passage on the supply ship M/S Johanna Kristina (called *umiarsuaq*, ship), travelling between Tasiilaq and each local village once a week.

Sermiligaaq

Around 80 km northeast from Tasiilaq, Sermiligaaq lies just south of the Arctic Circle (see Map 2.2 and Figure 2.3). It is the remotest of the settlements in the Ammassalik region. The village was established in 1922, yet, due to the people's semi-nomadic past, the wider area had been inhabited for centuries (Robert-Lamblin 1986: 87; Petersen 1984b). Sermiligaaq is located on a small peninsula at the edge of the Sermiligaaq Fjord, overlooked by a mountain whose slopes are popular for berry picking in the autumn. A short boat ride from the village through the interior of the Sermiligaaq Fjord, there are two big glaciers: the Karale and Knud Rasmussen. These glaciers gave the village its name Sermiligaaq, which means 'the place near the glacier'. In 2016, the population of Sermiligaaq numbered 197 people, with the population showing a slight decrease over the last few years (Kommuneqarfik Sermersooq 2016).

Sermiligaaq is known for its good hunting and fishing, and the village has a relatively high number of professional hunters and fisherman. Almost all male inhabitants go fishing and hunting in their leisure time. The area abounds in seals and a variety of fish species, with the ring seal being the most common seal species followed by the bearded seal, hooded seal, and harp seal. Common fish in the sea, rivers, and lakes around Sermiligaaq are: the arctic cod, char, Greenland halibut, redfish, capelin, and Greenland shark. Narwhals, beluga and mink whales, and occasionally walruses can be found in the waters. Apart from foxes and arctic hare, the only mammals found around Sermiligaaq are polar bears. In former times, hunting (mainly of seals) was the peoples' main activity whereas only a

Figure 2.3 Sermiligaaq, June 2005
(Photo: Sophie Elixhauser)

little fishing was done. Nowadays, the main income for hunters, who define themselves as both hunters and fishermen, stems from fishing. Seals are often caught
mainly for private consumption (and for use as dog food), and, to a lesser extent,
for commercial sale of skins, a source of income which has significantly decreased
throughout the last few centuries. This relates, among other things, to the considerable drop in the price for seal skins connected to anti-sealing campaigns carried
out by animal-rights activists in the 1980s and '90s (cf. Lynge 1992, for the situation in the Canadian Arctic see Wenzel 1991). These campaigns have drastically
impacted upon the economies of the smaller settlements in Greenland (Hovelsrud-
Broda 1997, 1999a, 1999b). Apart from hunting and fishing, the village offers a
number of employment posts, for the local administration and Greenland trade
(KNI), the school, and the kindergarten, amongst others. These posts are taken by
men and women alike.

Like most other villages in Greenland, Sermiligaaq has a basic infrastructure
consisting of a service house (*sullivi*), a power station, a meeting house (*katersutarfik*, sometimes called *forsamlingshus*, Dan.), a *Pilersuisoq* store (*pisiniarpik*)
located in the Greenland trade building – the latter also includes a room for bank
and postal services, and a selling point for helicopter and ship tickets – and a
modern school building (*alivarpik*). The service house is open on weekdays and
offers various services, such as laundry facilities, public showers that can be used
for a fee, a public kitchen, and a workshop as well as a room for processing skins.

The residential houses in Sermiligaaq are not connected to a water pipe system and only some households have water tanks which allow for running water. Members of the other households collect water with canisters from the water reservoir (*imertarpik*) in the middle of the village. The service house showers and laundry facilities are frequented often. The public kitchen is also a good place to get informed about the village news. The building also accommodates a small room for guests, which is rented out on a nightly basis. This is used either by tourists, who visit Sermiligaaq in the summer months, or by Greenlandic or Danish workers who have business to do in the village. The building of the *sullivi* also hosts the small health station (*niiarsiarpik*). The meetinghouse (*katersutarfik*) located directly opposite from the *sullivi* is an important location for the *inuusuttut* (teenagers and young unmarried adults) who gather there during evenings in order to play pool, table tennis, or the like. Dances take place at the *katersutarfik* twice a week, and sometimes community celebrations are held here. Celebrations may also take place in the school building, similar to assemblies of local associations and political communities. The school (*alivarpik*) serves educational purposes for children aged six to fourteen,[5] and every Sunday it hosts the religious service (Evangelical Lutheran).

Tasiilaq offers more employment opportunities than the villages. This provides an incentive for some young villagers to stay in town after finishing school (see Robert-Lamblin 1999b, for an account on social changes related to emigration from the villages in East Greenland). For Sermiligaaq, nevertheless, the problem of outmigration and young people not returning to the village after finishing school is not as striking as for some other small places in East Greenland, as Johan Uitsatikitseq, the Head of the Sermiligaaq school, has explained to me. Nevertheless, a general trend concerning outmigration from rural places to the urban centre Tasiilaq – which is quite similar in rural West Greenland – is for a higher number of young women to move to town, which leaves single bachelors behind (Perrot 1975; cf. Robert-Lamblin 1999b). Also in Sermiligaaq you find a few young hunters who have problems in finding a wife, which also relates to the fact that many young women prefer working in the supermarket, kindergarten, or one of the public institutions rather than opt for the hard physical work of being a hunter's wife.

Many Sermilingaarmeeq frequently travel between Sermiligaaq and Tasiilaq, especially during the summer months. Many are students attending secondary school in Tasiilaq, and spending holidays back with their families. Other villagers move back and forth between town and village because of jobs, boy- or girlfriends, family relations, or just because they want a change. In the summer, an important means of transport between Tasiilaq, Sermiligaaq, and the other villages in the district are small or middle-sized motorboats, which are mostly privately owned. During the summer boating season, Sermiligaaq can often become almost emptied of its residents, especially over the weekends. Many families leave for fishing trips, which are often combined with visiting relatives or friends in one of the other settlements. In the summer, moreover, many hunters leave for some weeks to more distant hunting areas, to return to the village only to stock up food or fuel. Sermiligaaq hunters habitually travel up north along the east coast to Kangerllugsuatsiaq and Kangerlussuaq for narwhal hunting, with women and children usually staying behind in the village. During the winter, when the fjords are frozen, mobility is limited and village life is quieter. Travel and visiting takes place less frequently between the settlements in the region, since the main (and costly) means of transport is the helicopter. A number of hunters in Sermiligaaq own sled dogs that are, above all, used for hunting and not in order to travel longer distances. Until recently, only one community-owned snow machine existed in Sermiligaaq, which is used for rubbish removal and other community tasks. In 2016, one of the Sermiligaaq hunters acquired the first two private snow-machines in the village, for his and his son's use.

The East Greenlandic language

Tunumiusut, the East Greenlandic language, is an unwritten Inuktituk dialect spoken by approximately 3,300 people along the Greenlandic east coast (added up by East Greenlanders living in West Greenland and Denmark as well as abroad). It differs to the West Greenlandic language phonologically and lexically yet the grammar is almost the same (Dorais 1981; Petersen 1975). Throughout the last century, Tunumiusut has variously borrowed from West Greenlandic (especially for words that hitherto did not exist), but it has not lost its specific character. Tunumiusut is not always understood by West Greenlanders (though sometimes it is, depending on context and person), whereas East Greenlanders do understand Kalaallisut (West Greenlandic), which is the official language of Greenland. School education in East Greenland is conducted in Kalaallisut, and all print media as well as radio and television programmes are in West Greenlandic. Yet, when the children start school, they first have to learn West Greenlandic before they are taught Danish or English. Compared to their counterparts in West Greenland, East Greenlandic children thus have to learn one more language, which contributes to the low educational level on the east coast.

Though officially all written communication takes place in Kalaallisut, East Greenlanders write their dialect in informal exchanges, mobile messages, and on Internet social networking sites, such as Facebook. The use of these media and particular ways of writing also impact everyday language (see Jacobsen 2006 for an examination of language use in Internet chat in West Greenland). The few printed documents in Tunumiusut (e.g., Maqe and Enel 1993a, 1993b; Maqe and Rosing 1994), and the handful of available word lists and dictionaries, use different orthographies (Gessain et al. 1986; Mennecier 1995; Robbe and Dorais 1986; Tersis 2008), whereas ways of writing among the inhabitants again differ from the linguists' orthographies.[6]

The lexical idiosyncrasy of the East Greenlandic language has sociocultural roots, as linguist Louis-Jacques Dorais (1981, 1984) contends. The most widely accepted explanation of the idiosyncratic nature of the language is the 'death taboo' (Dorais 1981: 46; cf. Holm 1911; Petersen 1975; Thalbitzer 1941). After a person's death, it was forbidden to utter his or her name until the name had been passed on to a newborn child, a practice which I still witnessed among some contemporary Sermilingaarmeeq. Likewise, if any object or being carried this particular name, it had to be renamed, which brought about a frequent modification in vocabulary. Dorais has argued, however, that this explanation does not account for all the changes in vocabulary, and that 'there seems to have existed . . . a more general tendency not to speak directly about potentially dangerous objects or forces (like animals and plants, the human body, hunting implements, etc.).' (1981: 46).

Evidently, language development and ways of speaking are closely embedded within the larger sociocultural frame, and are linked to ways in which people relate to each other and to the non-human environment more broadly. In order to understand interpersonal communication among the Tunumeeq, and the notion of personhood it entails, it is crucial to take a closer look at cosmological conceptions of the human person. Though nowadays not all inhabitants still remember the details with regard to these conceptions, their basic principles still underpin social interactions at large.

The human person

According to ethnographic sources, in the traditional cosmology of the East Greenlandic Iivit, in addition to the earth, there existed a land of the dead under the sea and another land of the dead up in the sky. The two lands of the dead seem to have been immaterial, Petersen writes, whilst the earth was material – yet populated by various non-human beings and entities (1984b: 631). The names for these beings have often been subsumed under the term spirits, which, as Holm writes, surrounded people everywhere but could only be seen by few initiated people (1911: 82). A human person was said to consist of three main parts, the body (*timi*), the soul (*tarneq*), and the name or name-soul (*aleq*) or name-soul, as well as a number of smaller souls (Petersen 1984b: 632). If one of the souls was removed, the body turned sick and, likewise, too much soul could make the person mentally deranged, or in extreme cases could also cause their death (Holm 1911: 80; Thalbitzer 1930). People followed various behavioural rules so as not to disturb the balance of the souls, which were more loosely connected to the body during various transitional phases of life, such as at birth, death, and at the onset of menstruation (Petersen 1984b: 632).

Souls could be stolen or negatively affected by people such as shamans (*angakkit*, sg. *angakkeq*) or sorcerers (*ilisiitsit*) who knew how to use different magical means such as *tupileq* magic. *Tupileq* magic conjured a being composed of parts of different animals, magically brought to life. It would be sent out to harm an enemy, but if the latter had greater powers than the person practising the magic, then it could be returned and affect the originator (Petersen 1984b: 632). Today, *tupileq* figures are sold as tourist souvenirs, however these are copies without any of the powers of original *tupilit* (Romalis 1983). An *ilisiitseq* practised evil magic only (Robbe 1983), whereas an *angakkeq* could interfere in beneficial as well as harmful ways. In case of disturbances in soul balance, an *angakkeq* could be asked to travel to the worlds of the dead to seek help from the so-called spirit world (Holm 1911; Thalbitzer 1930: 88–100; Robert-Lamblin 1996, 1997b; cf. Jakobsen 1999).[7] There were a number of mythical figures, sometimes also called deities in the literature, which a shaman could visit during his or her voyages, such as the sea woman (*imaq ukuua*), who is known as Sedna among other arctic peoples, the moon (Gessain 1978), or *asiaq*, a female figure associated with the winds and the weather (Robert-Lamblin 1996). A shaman could not only reinstate a balance of souls, but also incite a change in weather, and bring animals back to earth which had been retained by the sea woman responding to people's misconduct.[8] However, to obtain help from non-human powers one did not have to be a shaman, and there were a number of magical formulae and amulets that could be used by anyone (Petersen 1984b: 632–3; Robbe 1983).

Today, most East Greenlanders adhere to the Lutheran faith, and there is no shaman in the region, at least from what I was told. Tunumeeq I spoke with usually said that they did not know of anyone, or that 'maybe' (*uppa*) there is no shaman now. Knowledge of shamanism varied. Some inhabitants are still much aware of these beliefs, yet others seemed to know only a little. Names, however, are still important in East Greenlandic society, and it is in naming practices that beliefs in different souls still notably come to the fore. Information about naming practices in East Greenland is not only crucial when enquiring into the East Greenlandic concept of the person, but also in order to understand patterns of everyday communication and questions about who communicates with whom and how.

Naming practices

When a person dies, his or her name is passed on to a newborn, who, according to the Iivit, takes on characteristics and kinship relations of the deceased person, and in some ways becomes this person.[9] Formerly, as already noted, a deceased person's name (*aleq*) was not uttered until it had been given to a baby, an observation that was also followed when a person was thought to be in danger of dying (e.g., a hunter returning late from a trip) (Gessain 1980: 415). The practice of not pronouncing the name of a person after his or her death is still followed by some people (but not by everyone), and generally for family members kin terms are the preferred form of address. For instance, one day in 2007, an elderly woman in Sermiligaaq who was very ill was picked up with the ambulance helicopter. I talked about it with members of another family I often visited, and who quite obviously now avoided her personal name, which particularly struck my attention since they had previously spoken by name about my elderly friend.

The degree to which people observe namesake beliefs varies from context to context – sometimes namesakes are treated as other family members, whereas at other times it might merely imply that the specific kin term is used in addressing this person. Addressing namesakes with the kin term is widespread. My elderly friend Paula, for example, who passed away in 2009, used to call one of her granddaughters *anaana* (mother) and one of her grandsons *qalanngit* (sibling). I witnessed this practice frequently.

The passing-on of the name of a deceased person to a baby, or sometimes to several babies, takes place both within the wider kin group of the deceased, and beyond (though the former is more frequent). Most children receive several names, though usually one of the names is privileged (Gessain 1969: 243). Each name reactivates a set of relationships between the baby and the family of the deceased, who thus might 'see' a particular ancestor in this person. Accordingly, sometimes people are addressed differently by different people. A few times I became confused when a person I had come to know was spoken about using a different name. Persons also sometimes decide by themselves that they want to be called by another name, which can relate to a critical experience, such as an illness, family separation, or the desire to signal a change in their life (Robert-Lamblin 2000: 50; cf. Bodenhorn 1988: 11).

Names and naming are common topics of conservation (either with a serious voice, or in a jest), and people sometimes draw attention to namesakes' similarities in physical appearance or personal characteristics. This also applies to people from outside of the community. At times, I was told about other inhabitants of the region carrying my name, and a few times I was jokingly addressed by the kin term of such a person. A baby is only considered a proper human person (*iik*, pl. *iivit*) once it has received a name, which is nowadays usually proclaimed with baptism (cf. Nuttall 1993: 127). A person, and his or her reason and ability to think (*isima*), grows throughout life. Hence, the name, or name-soul, provides for the continuity between the living and the dead, and assures the continuity of personhood, which extends beyond an individual life.

Kinship and relatedness

Many authors writing about Greenland have stressed the importance of kinship, which entails a complex web of relationships including the living and the dead. Kinship in Greenland is not prescribed by consanguineal connection, as noted above, but encompasses consanguineal kin as much as namesakes and 'chosen' kin (see, for example, Nuttall 1992: ch. 6, 2000b; Robert-Lamblin 1999a; cf. Bodenhorn 2000a, on Inupiat in Alaska). In everyday life in East Greenland, kin bonds are very close, and a friend in Sermiligaaq once remarked, '*tamaani tamarmik ilarulaq*' (everyone here is related). This might also be influenced by the comparatively low population numbers as well as the relatively isolated location of the region. Most of the population in the region can be divided up into a number of large extended family groups (that can be distinguished when looking at the regional distribution of family names). Though nowadays these large family groups have been fragmented into smaller units, to some extent family networks still play an important role in everyday life, and a majority of daily interactions take place within the bonds of the extended family (Robert-Lamblin 1999b). Family membership entails various obligations and responsibilities, such as sharing (whose importance, however, has decreased considerably), or the obligation to accommodate relatives from other settlements for any period of time. It is common for East Greenlandic households to be made up of a large number of persons, and hospitality forms an integral part of social life.

As already noted when speaking about kinship relations with namesakes, kinship includes consanguineal as well as non-consanguineal relationships, which are inherently flexible and constantly created, re-negotiated, or, at times, cut.

Referring to a village in Northwest Greenland, Mark Nuttall has pointed out that '[t]he boundaries of kindred – or descent groups . . . are shifting constantly, as are the interpersonal relationships that are defined in terms of kinship' (2000b: 34). This claim also applies to East Greenland. 'Choosing kin' takes place through applying a kin term to someone who is not consanguineally related, who then becomes one's kin. Non-consanguineal relationships are considered just as 'real' as consanguineal ones, though people do recognise a difference. Nuttall speaks of an 'extensive improvisation in that people can choose their kin' (2000b: 34). Kinship, thus, is constituted through specific mutual behaviour, including obligations as well as rights. If you cease treating your relatives in the proper way, you may lose your kin status, despite consanguinity (cf. Bodenhorn 2000a). And these expectations of how to behave inevitably influence patterns of communication. This idea that particular behaviour and shared activity constitutes people as related resonates throughout hunter-gatherer ethnographies (Ingold 1999a: 406; cf. Bird-David 1994: 594; Myers 1986).

Adoption and fosterage

One type of this flexible kinship, which I frequently encountered throughout my fieldwork, is adoption and fosterage. In the majority of East Greenlandic households I became acquainted with I found at least one child who was not directly descended from the parents. Children in East Greenland play an important part in social life and receive a lot of attention. Early pregnancies are very common and young girls often have their first child in their teens (but the average age seems to be rising). The children of these young mothers are often brought up by the grandparents (or by other family members' families), which allows the young mothers to continue their teenage lifestyle and to finish their education. Marriage often takes place in one's mid or late twenties, and it is not unusual for young women to have several children before getting married (cf. Robert-Lamblin 1999b). Having had children from different men is not a hindering factor when searching for a partner, and young men commonly take up the role of a father for the children of their partner/wife, just as the other way around.

Though nowadays there is increasing recognition of the ideal of finishing education before having children, which is encouraged by a range of governmental institutions, this is often not put into practice, especially in rural places. Contraceptives are unpopular, and the number of abortions is high (see Emdal Navne 2008 for a discussion of reproductive decisions and motherhood). Various governmental campaigns try to encourage the use of contraceptives, through radio and television advertisements, or through school education programs.

Similar to other Inuit groups around the circumpolar North (Bodenhorn 1988; Guemple 1979; Mauss 1906: 468–9), it is common to give away a child for adoption or fosterage. Robert-Lamblin (1986: 72) distinguishes three different ways in which this can happen: 1) a child is abandoned by its mother; 2) it is 'given' by her; 3) it is asked for by the adopting parents. The first kind of adoption or fosterage detailed by Robert-Lamblin happens when social, material, or physical problems make it so that a mother is no longer able to take care of a child. Accordingly, a mother may decide to turn her baby over to the care of an institution, which will attempt to find a 'new' family. In some cases, however, it is not the mother but the institution which decides that a child does not receive the proper care. The foster families are often situated outside of the child's extended family. They receive a monthly allowance for taking care of the child, which at times influences their decisions. In many cases, various social problems lie at the heart of these cases of fosterage, sometimes turning into adoption, which influence the negative image of East Greenlandic adoption practices in Danish (and West Greenlandic) circles. This negative overall image, however, might also be influenced by the particular experiences of health and social workers who primarily deal with these 'problematic' cases.

The second type of adoption or fosterage, detailed by Robert-Lamblin, are the so-called 'gift children' – children 'given' by the mother. Mothers sometimes decide to give a child to a family member or friend who is not able to become pregnant herself, or who wants an additional child for one reason or another. Most East Greenlanders are very fond of children, and miss their company once their children have grown up and left. Thus, middle-aged or elderly couples are often keen to get a 'new' child to take care of, as are women who are infertile or who have only one child. East Greenlanders use the term for 'present' for this practice, such as by saying '*pitsavit*' (I give you a present). Agreements are sometimes reached pretty quickly, either already during the pregnancy, at the birth of the child, or shortly after. The given child then 'belongs' to the new family. This particular type of 'gift' is mainly exchanged between kin. The child will know about his or her consanguineal parents but not necessarily retain strong ties (Robert-Lamblin 1986: 73). Yet, frequent visits among family members in a particular settlement might make it so that child and consanguineal mother (as well as sometimes father) regularly see each other in daily life. Third, children are sometimes requested. For example, friends in Sermiligaaq would occasionally mention their wish to 'get' a child, and sometimes would approach a pregnant woman within the extended family about it. The latter might then decide to follow or not to follow this suggestion, and once again might be persuaded to do so. Robert-Lamblin notes that '[s]ometimes parents who have "given" one or two of their children "request" others a few years later' (*ibid*).

In the case of a number of families in Sermiligaaq I spent a lot of time with, it took me quite a while to figure out which child in which household has been born to whom, to the parents it lived with, to a sister, or possibly to another female relative. All in all, adoption and fosterage are perceived as positive practices among Tunumeeq, and, in most cases, I found that attitudes towards these adopted children did not greatly differ from the ways consanguineal children were treated (though usually everybody in the community was well aware of the child's consanguineal parents). Parents, just as Bodenhorn has written with respect to Inupiat, are those who have children, who do the parenting. Biological kinship is never denied, but the primary relationship, both in affect and in moral weight, are formed with those who brought you up . . . The actual physical reproduction of children is not as important as facilitating their social production.

(1988: 15)

Nevertheless, the frequent passing on of a child from one family to another, which most often happens in cases of fosterage/adoption occasioned by social or material problems, might at times negatively influence a child's development. This argument is often brought forward by institutions of social welfare and (often) Danish health workers (see e.g. Mejer 2007). I would argue, however, that these problems are due not to adoption practices as such, which have proved useful for many centuries throughout the circumpolar North, as still sometimes today, so much as to various underlying issues of social and cultural change.

The basic socio-cultural premises and themes described in this chapter, on a fundamental level underpin people's interpersonal encounters and patterns of everyday communication, which will be explored in the following chapters. Each of the chapters takes a particular social and/or spatial setting as its frame. As the East Greenlandic Iivit have been, and to some extent still are highly mobile, I will first elaborate upon communication connected to people's everyday travel.

Notes

1 There are, however, some indications that over the centuries a few individuals from Iceland/Norway have been stranded on the east coast. Also earlier trading contacts with the west coast of Greenland are reported (Robert-Lamblin 1986: 11). On the discovery of East Greenland see also Enel and Basbøll (1981).
2 See Krogh Andersen (2008), the book of a Danish journalist, which deals with the relations between Denmark and Greenland and various social, economic, and political issues in contemporary Greenlandic society, and also sheds some light on the position of East Greenland therein.
3 See also the memories of the East Greenlandic shaman Georg Qúpersimân (1982), edited by Otto Sandgreen.
4 These people are often either East Greenlanders with mixed Greenlandic-Danish background, West Greenlanders who have come to the east coast for work purposes, Danish inhabitants (and sporadically people of other foreign nationalities), or Greenlanders with Danish parents.
5 In 2006–7 the school employed five local teachers, all of them East Greenlanders, four of whom were *timelærere* (Dan.), i.e. teachers without a formal education paid on an hourly basis. In the recent past, every few years a Danish teacher supplements the local teachers and usually stays for one or sometimes two years at a time. During my fieldwork from summer 2006 to summer 2007 there was no Danish teacher in the village, yet shortly after my departure, a Danish teacher arrived and stayed for a period of two years.
6 For example, many East Greenlanders put an *r* at the end of a word, instead of a *q* (e.g., *piniartor* instead of *piniartoq*, for a hunter). For reasons of consistency, the orthography adapted in this thesis follows Robbe and Dorais (1986).

7 A shaman could be both male and female (Oosten 1986). For more information on shamanism in East Greenland see Jakobsen (1999); Robert-Lamblin (1996, 1997b); and Thalbitzer (1930).

8 On the myth of the sea woman in Greenland, see Maqe and Rosing (1994) and Sonne (1990), and for the eastern arctic version, see Laugrand and Oosten (2008).

9 Cf. Gessain (1980) and Robbe (1981). For similar practices among Inuit around the circumpolar North, see Alia (2007); Bodenhorn (1988, 2006); Guemple (1993); and MacDonald (2009).

3 Moving

Communication and everyday travel

Travel forms a significant part of people's life in East Greenland and is characterised by manifold encounters with other people, animals, and environmental features. Movements within the wider region take place between the different settlements in the Ammassalik area, and back and forth between hunting grounds, fishing locations, and camping sites. People make use of private motorboats, the supply ship, helicopters, snow scooters, and dog sleds. These journeys are very social, and through travelling together and meeting other East Greenlanders on the way, friendships or kinship relations are established or renewed. Moreover, through these encounters meaningful places are created and maintained, places that form part of the people's individual and social memory. This chapter takes a closer look at these encounters and relates communicative patterns whilst travelling to wider issues of community life, interpersonal relations, and personal autonomy. It builds upon the premise that movement is itself intrinsic to communication, and that one cannot separate out the movement of communication from the movement of people. Every conversation always involves the communicant moving about, be it through a gesture, turning head, or averted gaze, or through more obvious 'mobile' practices such as walking, travelling with the boat, or other means of transport. Hence, I regard communication as a joint movement of various co-travellers, who may be human or animal. Communication implies following the movements of others and reacting and adjusting one's own movements. This mutual engagement with and responding to the moving other, or in other words, the mutual attunement whilst moving *along*, distinguishes communication from mere perception. Communication, thus, does not happen on top of people's activities of moving, but as an aspect of them. As Vergunst (formerly Lee) and Ingold have described with respect to their research on walking,

> it is *through* the shared bodily engagement with the environment, the shared rhythm of walking, that social interaction takes place. People communicate through their posture in movement, involving their whole bodies. Crucially, walking side by side means that participants share virtually the same visual field. We could say that I see what you see as we go along together.
>
> (Lee and Ingold 2006: 79–80)

The landscape people pass through whilst being on the move is not merely a set-ting for communicative encounters or a background to human action. Landscape or seascape has to be regarded as 'the outcome of particular processes of engage-ment between people and the world in which they live' (van Dommelen 1999: 278). These issues have been discussed in various contributions from landscape studies, archaeology, and environmental anthropology, which question the sharp divide between the natural/material and human/social dimensions of landscape or environment (cf., Árnason et al. 2012; Hirsch and O'Hanlon 1995; Ingold 1993b, 2000; Tilley 1994).[1]

Processes of engagement between manifold beings and entities contribute to the creation of meaningful places in the East Greenlandic environment. Using the term 'place', I refer to Edward S. Casey, who defines place as 'the generatrix for the col-lection, as well as the recollection, of all that occurs in the lives of sentient beings, and even trajectories of inanimate things. Its power consists in gathering these lives and things, each with its own space and time, into one arena of common engage-ment' (1996: 26). Thus, places as arenas of engagement are intimately bound to sentient beings. Lived bodies belong to places and help to constitute them, just as lived bodies are always emplaced (Casey 1996: 26; cf. Merleau-Ponty 1962). Following Casey, places may not be set in contrast to an abstract, absolute space, as has been usual in much of the social science literature (cf. Whitridge 2004). Space and time, rather, arise together from the experience of place itself. Places, moreover and more broadly, gather. They 'gather things in their midst – where "things" connote carious animate and inanimate entities. Places also gather experi-ences and histories, even language and thought' (Casey 1996: 24). For example, Sermilingaarmeeq mentioning of place names was often accompanied with a story about past happenings at this location, such as in relation to particular journeys, people, and subsistence activities. Hence, places, and more broadly landscapes, embody history, or as Yvon Csonka writes, landscape *is* history (2005: 324; cf. Ingold 1993b, 2000; Nuttall 1991; Whitridge 2004). The embeddedness of history, landscape, travel, and people has been summarised by Ingold:

> It is the knowledge of the region, and with it the ability to situate one's current position within the historical context of journeys previously made – journeys to, from and around places – that distinguishes the countryman from the stranger.

(2000: 219)

Though nowadays not all East Greenlandic Iivit travel on a regular basis, as was the case in former times, being en route is still a highly appreciated condition and a theme people often deal with in everyday life. In the past, East Greenland-ers' semi-nomadic lifestyle implied movements within the wider region during major parts of the year. Travel was 'not a transitional activity between one place and another, but a way of being', as Claudio Aporta (2004: 13) has written with regard to Canadian Inuit. The contemporary situation in East Greenland, which is characterised by a settled lifestyle with obligatory school education and wage work, binds large parts of the population to the settlements during the week. Nonetheless, travelling is still widely practised and valued, though often confined to the evenings, weekends, or holidays. Exceptions in this regard are professional hunters and fishermen who spend much of their time en route out on the sea. Equally, those staying behind in the settlements, who are not travelling much by themselves, are also concerned with manifold issues of travel in their daily lives, be it through keeping up to date about other people's travel plans, through the flux of people arriving and leaving the village (or town), or through stories about former journeys (cf. Elixhauser 2015, forthcoming).

Building on my own extensive travel in the Ammassalik region, such as when accompanying Sermilingaarmeeq or other East Greenlandic friends, most examples presented stem from boat travel, the travel option that was most accessible to me (apart from helicopters). I will not talk much about dog sled trips, since travelling with dog teams during winter times is mainly a male activity which I could join only on a few occasions. Snow scooter travel is also excluded here since this means of transport is rarely used in Sermiligaaq.

Humans and animals

Much of the communication en route is related to animals (*ummasut*) in the surrounding waters, on land, or in the sky. East Greenlanders frequently call attention to animals they have caught sight of and comment upon these during particular activities such as hunting and fishing or whilst passing through their surroundings for other purposes.[2] People establish particular connections with animals and sometimes communicate with them in ways that parallel the ways humans interact with each other. Similar to Inuit elsewhere, according to the ethnographic record, the distinction between humans and animals did not exist in the mythical past in the way it does now. Referring to Inuit in Canada, Robert Rundstrom states: 'According to traditional Inuit environmental thought, men and animals were more closely related in the distant past, a time when they could actually speak to each other' (1990: 163). This also applies in East Greenland, as is evident from a great number of tales and legends, tales of animals in human appearance, and of non-human and human beings appearing as animals (see, for example, Holm and Petersen 1912 [1887]; Maqe and Rosing 1994; Rasmussen 1921; Thalbitzer 1930).[3] These and other ethnographic sources attest to a relationship between humans and animals based on respect and mutuality, going along with the belief that the animal gives itself up to the hunter (*piniartoq*), submitting itself so that the hunter may take it, just as the hunter shows respect in only taking those animals that willingly offer themselves (cf. Gessain 1969: 85; Robbe 1983: 37, 1994). This was, for example, expressed in ritual injunctions and customs in relation to hunting (Holm 1911: 49), only few of which seem to be known still or sometimes practised today. Grete Hovelsrud-Broda, who conducted fieldwork in Isertoq, East Greenland in the beginning of the 1990s, recounts a story of such sanction. In this case, however, it was imposed by so-called spirits and not by humans. The story was told by a local hunter:

> I came out the morning after I had caught a big polar bear, my thirteenth, and all his paws had been cut off. This was a sign from the spirits that I had now caught my last polar bear. When the spirits cut all paws off a bear during the night it means that a hunter will not survive if he hunts any more bears. I have not hunted a polar bear since.
>
> (Hovelsrud-Broda 2000a: 155)

Tunumeeq have not specifically talked to me about this understanding, but the behaviour towards animals I observed was also always characterised by respect.[4] Moreover, I could notice parallels between the people's communicative behaviour towards animals and towards other people, particularly with respect to vision.

Ann Fienup-Riordan (1986), who conducted research among indigenous people in Alaska, has written about the Yupiit understanding of personhood, which includes human beings as well as animals. She highlights the intrinsic power of vision, and recounts the tale of a boy who lived and travelled with the seals for a year in order to gain hunting knowledge and power. In the narrative the boy's host, the bearded seal, tells the boy to watch a good hunter: 'When his eyes see us, see how strong his vision is. When he sights us, our whole being will quake, and this is from his powerful gaze' (Fienup-Riordan 1986: 265). Additionally, he advises the boy never to look into the eyes of a woman, since then his vision will lose its power: 'you must be stingy with your vision, using it little by little, conserving it always' (*ibid*). Fienup-Riordan writes that the advice to beware of eye contact with women is well remembered by many contemporary Yupiit.

This closely parallels my observations and experiences with East Greenlandic Iivit. Here, as with the Yupiit, directly looking into the eyes of a person of the opposite sex bears sexual connotations. More generally, the particular power of vision, its meanings and practices, has become apparent in various situations of interpersonal communication. For example, there are many implicit rules on how and when to show curiosity and what a person is allowed to look at or ask about. Both in relation to curiosity and/or female-male communication, I noticed in some situations that East Greenlanders I was with always directed their gaze elsewhere, but not to the person or phenomenon being attended to. Likewise, cases of interpersonal dissonance or conflict were often expressed through not looking at this person and acting as if he or she was not present. Fienup-Riordan stresses that the particular power of gaze applies both to communication among humans and to human-animal communication, which seems very similar to East Greenland. She explains:

> this was much more than a matter of etiquette circumscribing his manner of vision within the human world. Rather, restricted human sight is profoundly significant – framing a man's future relationship with the seals and other animals, on whose good opinion, as a hunter, he would depend. A person's

vision, like his thought and breath, must not be squandered lest he be left wanting in his relations both within and beyond the natural world.

(1986: 265)

When I joined hunting activities, I was struck by the absolute concentration and tension inherent in the hunters' gaze when on the lookout for seals (*puilit*) or other animals (see Figure 3.1). After having returned from a hunting trip, which I will describe in more detail later, I wrote in my field diary (April 12, 2005):

> It is highly enthralling to watch Nikolaj's gaze, this concentration inherent in his visual search of the sea, absorbing every micro motion and reacting in a flash . . . His gaze glides across the sea, slowly, over and over again, going from left to right, from right to left. Then, all of a sudden, he catches sight of a seal. In the blink of an eye he grabs his gun and targets.

Hunters often follow the seals' movements, for instance under water, whilst driving side by side to them, or following them diagonally behind. Here, the seal and the hunter recognise each other's presence, or communicate with each other, yet not through seeing each other in a focal kind of way but through peripheral areas of the visual field.

Research from neurobiology has suggested that peripheral vision is especially sensitive to movement. In an article about Capoeira, a Brazilian martial arts and dance, Downey states that 'some researchers believe that peripheral areas of the visual field are more attuned to sudden movements than to detail for object identification or colour discrimination' (2007: 231; cf. Goodale and Milner 2004). Likewise, Goodale and Westwood argue that

> there is a wealth of psychophysical evidence that is consistent with the general view that in specific situations, particularly where rapid responses to visible targets are required, visuomotor control engages processing mechanisms that are quite different from those that underlie our conscious visual experience of the world.

(2004: 297)

It thus seems probable that for the hunter, just like for capoeristas and many sports practitioners, the key is to train peripheral vision, or the 'sideways glance', as Downey (2007) calls it. Hunters pick up movement from across a wide field, and detect animals as movements and not as objects. By not focusing on any particular thing, they keep the whole field of vision as sensitive as possible, which allows for rapid response. This contention is in line with the particular power of focal gaze among the Iivit that comes to the fore in different realms of everyday life.

Communication between humans and animals, which occurs in East Greenland as among Inuit throughout the circumpolar North, also involves mimicking the sounds of the animals. Robert Rundstrom, for instance, has written about Canadian Inuit ways of imitating animals.

Good hunters still knew the sounds to which seals, walrus, polar bears, and caribou would respond and frequently engaged them in conversation as they approached. The act was as much propitiation and appeasement of the soul of the highly respected animal as it was decoy.

(1990: 163)

Likewise, during a number of boat trips in East Greenland, young men were making sounds in order to strike up communication with birds, seals, or whales, trying to attract them. Thalbitzer stated, referring to East Greenland at the beginning of the 20th century, that '[m]any hours of a roving Eskimo hunter's life are devoted to repeated practice in these extreme articulations', and he argues that the frequent imitation of animals has come to influence Inuit language development (1974: 324; *in* Therrien 1987: 118). Rundstrom (1990: 163) mentions that kinesic, or gestural mimicry, frequently accompanied oral communication with anticipated prey, especially in pre-firearms times. This is also reported from East Greenland. For example, during spring sealing on the ice, seals were approached by mimicking their movements, as Thalbitzer reports, and I observed this a few times from a distance. Thalbitzer recounts, 'If the seal notices the approaching hunter, the latter

Figure 3.1 Seal hunting in the Sermilik Fjord, April 2005
(Photo: Sophie Elixhauser)

deceives it either by imitating the puffing sound and movement of a seal with the head and body whilst creeping forward, or by pushing a block of ice like a shield in front so that the seal may not see him' (1914c: 399). Hunters have frequently demonstrated these skills of imitating animals to me while telling hunting stories.

East Greenlanders' journeys not only involve communication *between* humans and animals, but also afford communication among humans *about* animals. Generally, in many situations, as, for example, during trips with the supply ship M/S Johanna Kristina, my fellow East Greenlandic passengers were very attentive towards the environment passing by, standing close by the window looking outside and pointing to specific land- or seamarks, or animals such as seals or whales they had spotted out in the sea. Also when travelling in small motorboats, my companions – both hunters and other passengers – would often search the sea with absolute concentration and then, on catching sight of an animal, would sometimes inform the other passengers through words, gestures, or mimicry.

The inhabitants know manifold gestures signifying animals (as well as means of transport, people, environmental features, and so forth). One of the gestures frequently used is an upward, outstretched fist signifying a seal (Figure 3.2). Tunumeeq, moreover, make use of a number of different hand signs for seals in different situations, such as an open hand palm facing upwards for a seal lying on the ice, or for the plural, both hands used in a similar manner but with moving fingers. There are gestural movements for narwhals (Figure 3.3), various other kinds of whale, and other animals (cf. drawings in Victor and Robert-Lamblin 1989). In many realms of everyday life, Greenlanders make use of a variety of gestures to denote particular beings, activities, or features of the environment.

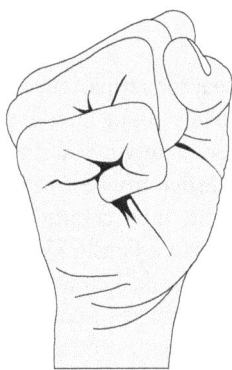

Figure 3.2 Gesture for a seal

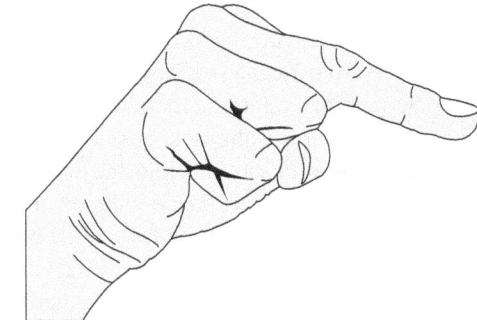

Figure 3.3 Gesture for a narwhal

Talking about animals whilst travelling is often linked to an exchange of information on locations abundant with fish or other animals. For example, in the summer months many Sermiligaaq families who have the possibility (i.e. access to a boat) leave by boat for fishing (and hunting)[5] on the weekends, or berry picking in the late summer. When joining friends we often met other boats out in the fjords, sometimes exchanging some words (or gestures) on our catch and possible fishing spots. Sometimes larger numbers of boats gathered at one of the particular fishing locations in the fjord near Sermiligaaq. Every settlement in the region has a few of these popular fishing (and hunting) spots nearby, frequented by the inhabitants. The exact locations of these places often change, adapting to the movements of the fish, and new fishing locations are usually quickly communicated to other boats nearby, often using gestures. Talking about animals is thus closely connected to the wider topic of wayfinding and orientation.

Communicating travel routes and orientation

Wayfinding among the Inuit depends on the people's knowledge and experience of the land- and seascape they travel in. It implies a multisensory monitoring of one's surroundings and, as Ingold has written, 'depends upon the attunement of the traveller's movement in response to the movements, in his or her surroundings, of other people, animals, the wind, celestial bodies, and so on' (2000: 242). A number of ethnographers working in different northern areas have expressed their amazement about the Inuit ability to find the way during periods of bad visibility, and have drawn attention to the senses involved in orientation (Carpenter 1973: 333–7; MacDonald 1998: 160–91). From my experience in East Greenland, not only did my companions rely heavily upon different senses to find their way through the sea (*imaq*) – such as touch, taste, hearing, or vision, just as Carpenter and MacDonald have described – but speech, different senses, and parts of the body were also highly relevant in communicative wayfinding with fellow boat passengers and other people encountered. For some initial impressions from my fieldwork, I will start with the description of a trip I undertook in late spring, 2007 (based upon my field diary, June 3, 2007).

> We were leaving Tasiilaq harbour in a small open motorboat for a trip of several hours from Tasiilaq to Sermiligaaq. Our group consisted of six young men from Sermiligaaq returning to their village after some time spent in town. Searching for a travel possibility to Sermiligaaq, just some hours earlier, my best friend from Sermiligaaq had told me on the telephone about the men commuting back to the village. Gedion, the owner of the boat, belongs to a family I knew well, and calling him on his mobile phone he agreed that I could join the trip. '*Tamatta aattartsaluut*' (We are leaving now), he told me. I hurried to prepare my luggage and proceeded to the harbour. Down at the harbour, the other passengers and I were waiting for Gedion and one of the other young Sermilingaarmeeq. The two came quite late carrying the fuel they had bought at the petrol station. Just this year a new petrol station selling fuel for boats and cars had

been built at Tasiilaq harbour. Some other East Greenlanders were hanging out at the dock with whom my companions discussed ice and weather conditions. Boat drivers and passengers just arriving at Tasiilaq harbour were questioned, as well as other East Greenlanders in town who had recently returned from a boat trip. People were telling us that there was much ice in the fjords, which makes boat navigation difficult. (During spring and autumn, the entrances to the three big fjords in the Ammassalik community – Sermilik, Kong Oscar, and Sermiligaaq fjord – are often blocked by huge chunks of ice.) In other situations, when arriving from a boat trip at Tasiilaq or Sermiligaaq harbour, I had been asked many times about routes and ice conditions: '*Suminngaanii?*' (Where are you coming from?) '*Sigekkaaju?*' (Is there much ice?)

We set off. The boat ride of 80 km takes several hours, depending on the strength of the motor, the weight of the boat, and environmental conditions. Our boat carried a lot of weight, with seven people, luggage, and fuel on board. At times I was worried it might be overloaded. We were not quite sure if our attempt to reach Sermiligaaq would be successful that day. The VHF radio set was turned on, and once in a while messages for other boats were received. Gedion and his cousin stood right next to the steering wheel, whilst paying attention to the radio. The messages sometimes contained important information on our travel route and environmental conditions. Shortly after disembarking we reached the trickiest part of our journey, the entrance to Kong Oscar Fjord right after Tasiilaq. As expected it was closed off by a large number of icebergs (*ililiat*) and ice floes (*kattit*). By chance, we encountered two other boats with Tunumeeq also trying to get through the ice. My companions and the people from the other boats exchanged some basic information on destinations and possible routes. Short questions and comments were shouted between the boats, yet more often directions were pointed out with everyone knowing what this meant. Gedion explained to me that the two other boats were trying to reach the villages of Kuummiut and Kulusuk.

At the outset, our group and the two other boats drove along the front of the icebergs searching for the most accessible entrance point. It did not look too good and, without a word being said, the possibility of having to turn back towards Tasiilaq felt tangible. After having checked the ice situation, our group and one of the other motorboats landed at the foot of a small island of rocks in the middle of the fjord entrance. We tied up our vessels next to another empty boat and two brightly coloured kayaks. The island is a common vantage point. Climbing up the rocks gave us a much better view of the situation of the ice. We were overlooking the wider region, mountains and fjords scattered with icebergs – appearing like white dots – and on the other side the open sea. The men had binoculars with them to search out possible routes through the ice. Up there we met some familiar faces from other settlements with whom my companions discussed the current ice situation. The small crowd was closely bunched as people looked around, sometimes pointing, gesticulating, or commenting. We also met four foreign kayak tourists who had hiked up the small, rocky island, with whom I chatted about possible kayak routes. My companions, all men in their early twenties, were quite curious about my English conversation with the tourists. Speaking only a few words of English, they did not understand what we were discussing. They stood within earshot. Later on Gedion asked what we had been talking about. '*Kisi orarpa?*' (What did they say?), and I explained what the tourists had told me about their projected kayak route.

Climbing down again and getting back into the boat, one of the other vessels found an entry point into the floating icebergs. It opened up a small trail within the frozen water for us to follow behind. There were three boats following each other one after the other, most of the time staying in viewing distance. Our boat was the last one. Using the boat like an icebreaker, we navigated through the ice-covered waters. This was a challenge, since the passage cleared by the boats ahead of us quickly froze over. The boats ahead did not take the followers much into account. Most of the time we followed the trail broken by the other boats; sometimes we found another, easier passageway. The ice was constantly moving, and within a very short time the ice cover could change; passages were getting blocked or becoming more accessible. Every now and again one of my companions used a lance (hunting tool) to push aside some of the icebergs so that we could pass between them. They did not talk much about what to do or which route to choose, apart from once in a while gesticulating or dropping a few words. Everybody seemed to know, and actions and reactions occurred without discussion. At one point we all climbed out of the boat onto the ice to lighten the boat's weight. The men

Figure 3.4 Pulling the boat over an ice flow, June 2007
(Photo: Sophie Elixhauser)

pushed the boat over a large ice floe back into the open water. After an hour of straining work, all three boats had passed the difficult area and drove off in different directions. We had reached the open sea.

We sped up. The wind made the temperatures much colder, and we put on warmer clothes. Our boat did not have a cabin or a little plastic roof on one side of it to protect against the cold as some of the other boats in the region do. Now the journey was getting much smoother. Gedion turned on some pop music on his new mobile phone. The music from Gedion's phone was not carried far; the sound of the boat engine was too loud. The rest of us were wrapped up with clothes. The young men were wearing the lined, colourful overalls that you can buy in the few shops in East Greenland, and which are commonly used by men around Greenland for all sorts of trips. We had found comfortable positions in the small boat, which was pretty packed with people and luggage. Two of my companions had a bottle of alcohol with them, and I noticed their behaviour slowly changing. I was asked more and more questions (e.g., about Germany and why I did not want to choose a boyfriend from amongst them).

We had turned into another fjord. We passed by the remnants of the former American military base Ikkatteq, and opposite a prominent mountain with red stripes. A few times Sermiligaaq friends had mentioned the beauty of this mountain. Our boat suddenly slowed down. The fuel was running out. Would we make it to Sermiligaaq, only some miles ahead of us? After a while, two other boats overtook us. One of them stopped next to us; it was my companion Tobias's parents returning from a weekend out fishing in Kuummiut and Kulusuk. Tobias's mother Kamilla is a friend of mine, and she waved over, signalling that I should join them on their much faster vessel for the last twenty minutes of our journey. Tobias also jumped into his parents' boat. We took off some clothes; this boat had a cabin and was much warmer inside. We exchanged some words about Kamilla and Massanti's fishing trip, and they showed me their sacks with fish they had caught. They had been staying with relatives and seemed to have had great times out fishing. Yet everybody was tired, and for the rest of the trip we did not speak much. When we arrived in Sermiligaaq, few people were outside at the harbour. It was already getting dark. I walked to my host family's house. Later that evening Gedion's boat also made it back to Sermiligaaq.

With many Sermilingaarmeeq travelling back and forth between the settlements in the region, boat pools are often formed using all the available space in a boat. Asking other villagers to join in a boat ride is a cheap possibility for travel. In the small region with dense social networks, word on who plans to travel where is quickly spread, and it is not difficult to keep track of possibilities for boat rides. Thus, as in the case of this trip, you often see boats with a large number of passengers on the move together. Usually, additional passengers pay a share of the fuel. Thus, wayfinding is often preceded by processes of 'boat-finding'.

Apart from the communities formed on boats and ships during travel, hunters and fishermen spend a great deal of their time out in the fjords, and often encounter other hunters or travellers during their activities. Hunters are habitually on tour either by themselves or in a team of two or three (or sometimes more). Hunting and fishing parties, in contrast to groups of travellers commuting between different settlements, are usually made up of regular hunting/fishing partners who are most often relatives (and male). For example, Mathias, a professional hunter and fisherman and the husband of my good friend Hansigne, is often joined by his two brothers who also live in Sermiligaaq. In contrast to Mathias, the two brothers have employment in the village and can only go hunting or fishing in the evenings, weekends, and holidays. Mathias therefore also frequently sets out by himself. Another one of Mathias's regular hunting partners is Rasmus, his brother-in-law, who lived and worked as a carpenter in Tasiilaq during 2006–7 yet spent holidays with his family in Sermiligaaq. On my visit in 2008 he had moved to Sermiligaaq to work as a full-time hunter, which strengthened the hunting partnership with Mathias. Professional hunters and fishermen usually have their own boats. Hunters and fishermen who do not own a boat are dependent upon relatives and friends to be taken along. The expectation to contribute to the fuel seems to be widespread, whereas the additional person may keep hold of his (or her) own catch. Moreover, sometimes the boat of a relative or friend is used on a longer-term basis, which is reciprocated in one way or another. The husband from my host family, for example, lent his motorboat to his brother-in-law for some months in 2007. In return, the latter gave him an unusually big share of his catch from hunting and fishing trips during that time.

Encounters during hunting and fishing trips may result in shorter or longer halts, or just in noticing each other from a distance. Being far away and out of sight, Greenlandic hunters I have joined were never surprised to meet another boat appearing out of the fog or from behind an ice floe. My observations parallel those of Nooter, who writes: 'It is evident that in many cases hunters paddling between the tall ice floes are well aware of the presence of other hunters some distance away' (1976: 41). Accordingly, when problems arise, such as the breakdown of an engine or difficulties in finding one's way through the ice, help can often be sought quite quickly. The awareness of the presence of others is created through various communicational means, such as direct encounters, recognising others from a distance or from a lookout, sometimes using binoculars, or through communication via the VHF radio.

Travel routes and orientation are important topics of conversation, both among the passengers of a boat (or dog sled, snowmobile) and with other East Greenlanders encountered on the way. Briefly stopping during travel (or before starting a trip), the passengers exchange destinations and origin of travel, and sometimes routes and environmental conditions. Many of the encounters on the move take place along memorised trails and routes crisscrossing the wider region that the people inhabit. Routes used by local travellers exist for dog sled travel in the winter times and boat travel during the ice-free months of the year.[6] These routes form networks of exchange and communication. Having been used for a long time, they are part of the social and individual memory of the inhabitants (cf. Aporta 2004). People meet whilst moving along these trails, during activities at various hunting and fishing spots, or whilst taking a break at popular stopovers or lookouts such as the light tower at the entrance of the Kong Oscar Fjord near Tasiilaq, or the small rocky island I described in the above account.

Thus, these routes of a traveller or a group of travellers get entangled with the routes of other inhabitants, or 'lines' as Ingold (2007) would put it. These entanglements imply the creation (and maintenance) of meaningful places – as for instance the aforementioned lookouts, the beautiful mountain that East Greenlanders often talk about and use as landmark, or one of the places that serve as meeting points during journeys – or are important for communicating travel routes or wayfinding more generally. As Casey writes, 'space (in the form of "definite trade routes") and times (in terms of "periodical journeys") come together in place – in places of exchange connected by regional pathways' (1996: 42). Yet the boundaries of these places are inherently porous and flexible, or to use Casey's words again, 'Places are at once elastic – for example, in regard to their outer edges and internal paths – and yet sufficiently coherent to be considered as the *same* (hence to be remembered, returned to, etc.) as well as to be classified as places of certain *types* (e.g., home-place, workplace, visiting place)' (*ibid*). Ingold calls these places knots; the knots come into being and are re-created (or disappear again) through people's practices, activities, and memory. He writes:

> The lines are bound together in the knot, but they are not bound by it. To the contrary they trail beyond it, only to become caught up with other lines in other knots. Together they make up what I have called a meshwork, and the threads from which it is traced are lines of wayfaring.
>
> (Ingold 2007: 100)

Accordingly, these places or knots of encounter in the sea- and landscape of East Greenland change and adapt according to people's travel patterns and are created and influenced by a variety of factors such as the different seasons, changes in weather, ice conditions, availability of prey, people's means of transport, or decisions on destinies and routes, to mention only a few. Some of these places are and have been frequented for a long time, have become significant markers for travel, and form part of people's oral history. Others, however, are relatively new, while still others have lost their importance and are no longer frequented or remembered.

I now want to look more closely at inhabitants' communicative patterns at these places of encounter and along these routes of travel, especially with regard to orientation and wayfinding. As my observations have illustrated, verbal indicators of directions and place names used by communities of travel or other people encountered on tour, are usually accompanied by gestures and facial expressions (see Figure 3.6). Sometimes the communication takes place mainly through nonverbal means. In addition, technologies such as the radio are used to communicate issues relevant for travel. Boat driving (as well as driving dog sleds) and orientation whilst travelling are men's tasks, and therefore women are involved to a much lesser extent in discussions and exchanges on wayfinding and travel routes. Women do join boat trips on a regular basis, but they do not take up the driver's position, and rely on the men for orientation.[7]

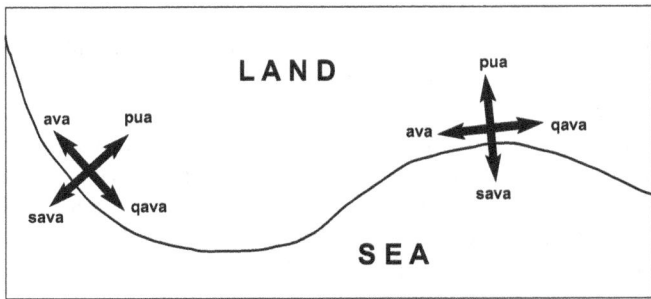

Figure 3.5 Orientation in East Greenland
(Illustration: Martl Jung, based on Robbe 1977: 74)

When talking about directions, East Greenlanders, like most hunter-gatherers, do not refer to an independent set of variables such as the points on a compass (cf. Fowler and Turner 1999: 424). They use two sets of terms to assign directions, both of which are aligned to the surrounding environment. The first and most common set consists of four directions, indicated by the terms *sava, pua, qava,* and *ava*, plus varying suffixes, which are defined according to the direction of the coast stretching out to infinity (see Figure 3.5, cf. Thalbitzer 1930: 85). The second set of terms is only used within a restricted area such as a fjord. Here, also, four directions are named – *kita, kangia, orqua,* and *kiala* (plus suffixes) – that refer to the mouth, the inward end, and the different sides of the fjord (Robbe 1977; see also Fortescue 1988, comparing orientation among different Inuit groups).

Moreover, pointing gestures and bodily movements play a prominent role in wayfinding and orientation, such as gestures showing directions or calling someone's attention to relevant environmental features. Apart from arm movements, these 'wayshowing' procedures (Mollerup 2006) also include movements with the head and facial expressions. These are not always accompanied by place names or verbal indicators of direction. When pointing towards a specific direction, for example, other Tunumeeq often knew where the person or group was heading without verbal specification, as illustrated in the trip with the young Sermilingaar-meeq I described. The passengers of the other boats we encountered did not name the villages to which they were heading; yet their gestures and arm movements were well understood by my companions. In this context, people often additionally rely on indicators or clues such as what kind of people are travelling, the luggage on board, the time of the day, and so on. For instance, assuming that a boat is heading for a settlement, given that there are so few settlements in the region, pointing in a certain direction is often easily understandable.

Moreover, gestures during processes of wayfinding are used to call attention to obstacles or other occurrences relevant for orienteering. They may be passageways, physical barriers, other boats approaching, animals in the vicinity, and so on. When I asked about wayfinding procedures whilst travelling with East Greenlandic friends,

Figure 3.6 Way showing during a fishing trip, Ikkatteq Fjord, September 2006
(Photo: Sophie Elixhauser)

the explanations given to me were often accompanied by movements of the arm, and sometimes comprised gestures alone. Having my attention drawn to something, an iceberg blocking the continuation of a particular travel route for example, it was often my task to understand the significance and implication of what was being shown to me. Extensive verbal elaborations were rare.

The general atmosphere often did not invite questioning or extensive talking. Travelling among East Greenlanders habitually includes long periods of silence. Yet these silences are often 'communicative' (cf. Jaworski 1993; Tannen and Saville-Troike 1985), and have to be seen as a sign of respect and of the welcoming acceptance of the other. Indeed, sitting together not talking can be a very communicative moment. As Frank Sejersen writes, this 'active silence' correlates with a way of talking shared by many Greenlanders that has to be distinguished from the kind of silence that conveys a lack of interest (2002: 68). Moreover, relating to East Greenland, Robert Gessain has mentioned the importance of respecting the 'silence of nature', especially crucial for hunters (1969: 140; Kwon [1998] reports similarly for the Siberian Orochon). Accordingly, though during the initial months of my fieldwork I often felt like asking for explanations, I learned to be patient and to wait for the right moments to find out more. Too much talking and questioning is often not appreciated by Tunumeeq, and contexts and topics, about which it is appropriate to ask questions, as well as appropriate moments in time to ask, differ from conventions I am familiar with in other European societies.

Returning to the topic of pointing gestures, I want to refer to the work of Andreas Roepstorff, an anthropologist who has made similar observations whilst navigating ice-covered waters in west-central Greenland (2007: 192). He recounts a boat journey together with a Greenlandic father and his son, during which the father nodded or pointed towards a specific direction, which featured something to which the boy should pay attention. Without further explanation, the boy then had to find out for himself what was being pointed out, and to take it into account in his planning of the way through the water. Through an 'education of attention' (Gibson 1979; cf. discussions in Ingold 2000), the boy learned to see specific phenomena that he formerly had not recognised or found important. Here we are dealing with a process of visual enskillment, of 'learning to see' (cf. Grasseni 2004, 2007b; on enskillment in the Icelandic fisheries, see Pálsson 1994). Nuttall likewise remarks that becoming a hunter in Greenland is the result of 'situated learning', or 'learning in practice', to understand the territory, the movements, and other habits of seals, and 'the hunter's place in the wider social context' (2000a: 42).

These descriptions relate to some overall characteristics of children's education and processes of learning in East Greenland. Children are often asked to pay attention to a procedure they are meant to learn, and take over the role of a silent, patient observer of the adults' way of doing things. Later on, children practise by themselves, often unsupervised. In many cases the results are shown to the adults only if successful. Verbal instructions are rare (cf. Briggs 1991, for a similar account for Canadian Inuit). I have encountered these characteristics of educating children not only through living in a family with many children in Sermiligaaq, but also through my own role of being an anthropologist learning the language and many details of daily life in East Greenlandic society, similar to the way a child learns.

Moreover, the frequent use of gestures also implies that the people communicating often look in the same direction, sharing the same visual field (cf. Lee and Ingold 2006). This applies not only to communication among boat companions, but also to communication between passengers of different boats. As explained above, travellers are often aware of other boats in the vicinity, be they travel companions or just some other boat one has met on the way; communication between different boats is often characterised less by face-to-face confrontation than by driving *with* the other, side by side or after each other, and by attuning movements to the other, which, just as during hunting, implies attending to peripheral areas in the visual field.

Wayfinding and leadership

Turning to a slightly different perspective, I will now relate the communication that accompanies wayfinding to leadership (cf. Elixhauser in press-a, in press-b). During the spring or autumn, when the fjords are still partly ice-covered, boats meeting on their way and heading in the same direction often try to keep together, especially around locations which are difficult to navigate. Usually, the most experienced boat driver goes ahead to form a path through the ice-covered waters that the other boats follow behind. This happened, for example, during the trip with Gedion and the other young men, in which our boat met up with two other boats at the entrance of Kong Oscar Fjord. For this difficult passage, one of the other boats took the lead, and our boat and another one followed behind. Usually the selection of who goes first happens smoothly and in silence, depending on the skills, experience, and technical equipment of the people present. In the case of the trip described, my companions in their early twenties had only little experience of wayfinding, while the driver of the boat going ahead had greater navigational skills. The three boats were of similar size and equipment. The leading boat managed to enter the icy passage first, and we others took the chance of an open trail to follow after. Which boat would take the lead had not been much discussed; instead, it had depended upon the personal initiative of that specific boat driver at that moment. For more illustration of wayfinding procedures, and the topic of leadership in wayfinding in particular, I will recount the events of another trip, which took place during my first fieldtrip in 2005 (field diary, April 7–11, 2005).

One day this spring, shortly after my arrival in East Greenland, I was able to join a hunting trip of several days with Nikolaj, a hunter from Tasiilaq/Tiilerilaaq. Nikolaj frequently travels throughout the region and visits the different settlements. His mobile lifestyle is influenced both by his job as a hunter and fisherman, and through his employment at the Red House Hotel, for which he often takes tourists for trips around the region (by boat or sometimes dog sled). This job provides him with the financial means for modern technical equipment – for instance, his boat is equipped with a very strong engine. During this first stay in East Greenland in 2005, I volunteered for the Red House and was sometimes able to join one of Nikolaj's trips, such as on this occasion. We were heading towards the Sermilik Fjord, and planned to sleep for some nights at the village Tiniteqilaq (or Tiilerilaaq, as the village is called in East Greenlandic). Our company included my friend Anita, another foreigner like me working for the Red House, and a young man from Tasiilaq, who had needed a lift to Tiilerilaaq. Nikolaj was among the first hunters that year to navigate his boat all the way from Tasiilaq to the Sermilik Fjord. At some locations, especially where the fjords lead into the open sea, the water was still densely covered by ice floes and icebergs. Only with much effort did we manage to pass through the mouth of Kong Oscar Fjord. This endeavour required all the available power of the motor as well as physical strength in order to push aside pieces of ice with a lance or by hand. After turning into the Sermilik Fjord, the waters were mostly ice-free. We had managed the most difficult segment of our trip and were now able to move about without problems. The rest of this first day we spent in the Sermilik Fjord seal hunting. Once in a while we met other boats with hunters from Tiilerilaaq. Some of these meetings were arranged via the VHF radio, which was running most of the time. But often we met other travellers by chance, along well-known routes or at locations favourable for hunting. After many hours out at sea and with two seals on board that Nikolaj had caught, we started heading towards Tiilerilaaq where we planned to spend the night. It had turned cold and foggy. The ice was getting denser, and navigation was more difficult than during the day. After some manoeuvring, with fog impairing visibility, we reached Tiilerilaaq. It was dark.

The next day we also spent out hunting. During some parts of the day, one of the local hunters joined our company. Nikolaj caught a great number of seals and in the evening our boat returned carrying eleven seals. This was very exceptional. Nikolaj is known as a great hunter, a *piniartorssuaq*. For the night, Nikolaj buried the seals in the snow at Tiilerilaaq harbour. The next morning we tried to drive back to Tasiilaq. Some of the local hunters helped Nikolaj dig out the seals and to put them back into the boat. The first part of the journey went without problems. But on reaching the problematic part around the entrance of Kong Oscar Fjord, we encountered more ice than expected. At one point we could go no further. We landed on a big ice floe and climbed up an iceberg (see Figure 3.7). Nikolaj explored the surroundings with his binoculars. He figured that there was no chance for us to drive back to Tasiilaq that day. We had to return to Tiilerilaaq, and wait for the ice situation to improve. It was a bright, sunny day and we stayed up on that iceberg for a little while relaxing and enjoying the sun. After returning to the boat, however, we had trouble getting the engine to work again. The propeller was

Figure 3.7 Outlook from on top of an iceberg, April 2005
(Photo: Sophie Elixhauser)

blocked by chunks of ice. Revving the engine and with the help of a harpoon-like tool, Nikolaj managed to remove the ice.

Not far from the blocked passageway, we stopped at a small rocky island where Nikolaj wanted to leave the seals, with a view to picking them up on the return journey the next time. At the foot of the island he skinned the seals and removed the fins and the claws, the most valuable parts of the animals. We all helped to dig holes in the snow. After burying the seals, we headed back into the Sermilik Fjord, and spent the rest of the day out in the fjord. We met other boats, socialised, and enjoyed the nice weather. Nikolaj did not catch any more seals that day. Dusk offered a great evening ambience with icebergs in front of a red sky and moving wafts of mist between them. Once in a while we saw another boat from a distance. We started heading back to Tiilerilaaq. The ice was getting denser and denser, and Nikolaj manoeuvred the boat between the ice floes. Time passed, and I had long since lost my sense of orientation. The temperature had dropped significantly. Visibility was low and the ice was freezing quickly. Our boat acted like an ice-breaker, and we needed its entire weight to break through the floes. After a while, some of the other boats we had met earlier in the day reappeared and silently fell into line behind our boat. Nikolaj stood behind the visor of his open motorboat, fully concentrating on the changing ice formations. His gaze focused on the surroundings, and at times, when driving over some layers of ice, or breaking a path through the ice, his movements and the movements of the boat seemed to be one. The others followed our trail. Nikolaj did not look back at our followers, nor did he take them into account in any apparent way. Already in previous days, I had noticed the respect paid to him by the other hunters. Nikolaj is the expert when it comes to difficult navigational situations, as I was later told by other East Greenlanders.

In this incident of joining an experienced hunter and boat driver, our boat was in the position of leading other boats through the ice. Nikolaj, thus, temporarily took over a leadership position. By and large, East Greenlandic society has long been characterised by a lack of formally organised leadership patterns, and the early ethnographic sources describe a kind of situational leadership (Campbell Hughes 1958: 369). Anthropologist Robert Petersen describes East Greenlandic society before the 1950s as 'cemented not by leaders but by the reciprocity of free hunters' (1984b: 639). Likewise, relying on fieldwork material from the beginning of the 20th century, Thalbitzer writes that Ammassalik society

knows neither lawfully recognized chiefs nor representative institutions. Only as an exception have there existed Eskimo chiefs, of whom tradition relates. It was a temporary occurrence, not hereditary condition . . . As a rule the hunter is head only of his own family, and has no authority over other families in the village. Custom, however, gives the oldest sealer in the village, or in the house, a certain degree of patriarchal authority, but this does not extend beyond the boundary of the village.

(1941: 618)

With colonization, and the integration of Greenland into the Danish state, various formal leadership positions have developed in politics, the schools, and businesses and institutions. The contemporary situation reveals both situational as well as formalised leaders, and the continuing importance of situational leadership in many arenas of everyday life influences the practices of formalised leaders.

Gert Nooter, a Dutch anthropologist who conducted fieldwork in Tiilerilaaq in the 1960s and '70s, wrote a valuable book about leader- and headship and changing authority patterns in East Greenland (Nooter 1976). Nooter distinguishes the operation of situation-specific, shared-goal leadership from that of quasi-institutionalised, imposed-goal leadership. The East Greenlandic language does not offer an overall term for leader, akin to *umialik* in the Western Arctic (Alaska), or *isugumataq/ ihumatar/isumataq* in the Canadian Arctic (*ibid*: 7–9). The terms used among East Greenlanders are *piniartorssuaq* for a great hunter (as, for example, Nikolaj) and for situational leadership in tasks widely related to hunting such as wayfinding and boat navigation, or *naalanngaq* (Nooter writes *nalagkak*), which denotes a leader in an official institutionalised context. This distinction between situational and institutionalised leadership was also much in evidence during my fieldwork. The situation-specific, shared-goal leadership is the kind we find in situations of wayfinding as described above. Another example is breaking a trail for the first time in the year, which is always led by an experienced hunter (cf. Aporta's account of Canadian Inuit, 2004: 18). Also the shamans (*angakkit*) were situational leaders in that their knowledge and practices in the shamanic realm were not tied to a position of influence and power in other areas of daily life. Sonne explains: 'The men tacitly followed the advice of the oldest, most experienced hunter in matters of fishing and hunting, and the *angákoq* had no power except in religious matters bearing on the existence of the whole community' (1982: 24).

By and large, someone becomes a situational leader because he or she has demonstrated ability in some spheres or has the best technical equipment, and many individuals have chosen to follow his or her suggestions. The leading position, which usually does not relate to other spheres of daily life, can easily be taken over by someone else the next time, if they perform better. Of course, some individuals are well known for particular skills and regularly take the lead. Yet, the choice to follow somebody is always voluntary, and giving orders or trying to convince others is not common, as I have found in a variety of contexts of daily life in East Greenland. Giving orders or telling somebody what to do is perceived as an improper intrusion into the other person's personal autonomy. Similarly, a situational leader usually proposes a way of doing which leaves the decision of whether to follow or not to follow to others. A skilled boat driver, for example, navigates the way he thinks is best, and other drivers decide if they want to follow. I was often told '*nammeq*' (that's up to you) when asking others about what to do. Accordingly, 'power works by attraction rather than coercion', as Ingold has stated for hunting and gathering societies more broadly, and 'the relationship between leader and follower is based not on domination but on trust', which is conditional upon respecting the followers' autonomy (1999b: 404).[8]

Similar practices of leader- and followership are apparent in many realms of East Greenlandic society, from the education of children to various personal decisions taken in daily life. In situations of wayfinding as well as in other comparable situations, both positions are situational and voluntary: the leading position, based on initiative, skill, and equipment, and the position of the follower, who decides for each moment if he or she wants to follow or not. Here, we find many parallels with what has been observed in other formerly egalitarian and non-hierarchical societies such as the Taicho Dene (Legat 2012) or the Naskapi (Henriksen 1993) in Canada.

Likewise, communication between the leading party and followers is characterised not so much by face-to-face exchanges, as by recognising each other whilst driving *along*. The driver of the leading boat, just as his fellow passengers, usually does not look back, though he is aware of the presence of the boat(s) following, perceived through the corners of his eyes. Learning happens whilst following, through enskillment among followers who are not being instructed. Processes of learning, and of leading and following are thus to a large extent based on peripheral vision, just as I have shown in my descriptions of hunting and wayfinding above. Communication in these situations entails noticing the movements of the other, be it an animal, or a boat following, and mutually engaging with and reacting to the other's movements.

Orientation, stories, and teasing

Though not directly connected to the assignment of positions of leadership in other realms of everyday life, success in wayfinding (as well as in hunting) influences a person's status and esteem in the community. In this respect, stories told at a later point, such as back in the village or town, play an important role. I will now turn briefly to a slightly different perspective and explore some aspects of communication *about* moving and travel, in particular related to wayfinding, including teasing and gossiping.

Skills in wayfinding and orientation are highly valued in East Greenland, as among other Inuit groups (cf. MacDonald 1998: 161). They can be a source of pride, but at times can also become a source of embarrassment (cf. Aporta 2004: 32 on Canadian Inuit). Stories of wayfinding under extreme conditions feature particularly prominently among Sermilingaarmeeq and other East Greenlanders. These stories may not only be rewarding for a boat driver or another member of the boat crew, but can also have a rather negative effect. For example, I have often heard Sermilingaarmeeq gossiping about hunters with poor performance or a person who could not find the way. After villagers had returned from a difficult trip, and for example had become lost, the details of these incidents quickly spread among villagers. I will give an example.

One day in December 2006, I was spending the evening at an elderly couple's home in Sermiligaaq. The man was recounting his hunting trip from the day before. He had been hunting with a relative of his, and at some point out at sea they had encountered his son-in-law standing by himself on an ice floe. It turned out that the young man's companions, his two cousins, had left him back on the ice floe for seal hunting, and had not returned to pick him up. I did not quite understand whether the cousins had just forgotten about him or if they had not been able to find their way back to the ice floe. When telling the story, however, it became quite apparent that the man and his wife, who had joined in the conversation, were criticising the two cousins' poor memory of how to return to the place where they had left the old man's son-in-law. They did not specifically say so, but the tone of voice, laughter, as well as the kinds of looks they were giving each other and to me and the other guests, made that very clear.

Another example of a story which was frequently recounted was an incident in spring 2006, in which I accompanied a group of friends from Sermiligaaq on a narwhal hunting trip to the Sermilik Fjord. We stayed out camping along the fjord for some nights. The first night we had put up our tents way back at the mouth of the fjord, or *paornakaiiq*, as the location is called in East Greenlandic. My friend Hansigne, her three-year-old daughter, and I (i.e. the women in the group) had just gone to sleep when suddenly the men who were still sitting in front of the tents called agitatedly from the outside. Huge amounts of ice were drifting towards us, they were saying, and we had to leave. Very quickly we jumped into the boats and drove off. We did not have the time to get properly dressed, or to pack our luggage. During the boat ride, the little girl Dina and I used the unfolded tent as a cover against the cold air. In the rush our warm clothes had become stored at an inaccessible location in the boat. A lot of ice drifted towards us, and navigation was quite difficult. We only just managed to leave that dangerous location, and after some hours of driving in the semi-darkness we reached another, safer location where we ended up spending the rest of the night. As my friends and companions told me later on, the ice could have locked us up at this location, and in the worst case our boat could have been blocked for several weeks. Yet at that moment I had not quite understood the seriousness of the incident. Later on, back in Sermiligaaq, I was teased about my naivety, as I had even once laughed after the men told us to hurry up.

Teasing, shaming, and derisive talk, and likewise gossip, are very widespread amongst East Greenlanders, and they range from joint amusement and laughter without much seriousness to voicing differences of opinion, or sanctioning behaviour (cf. Robert-Lamblin 1986: 145). Regarding the term *piniartorssuaq* (big hunter), for example, Nooter mentions its use in sarcasm, such as for a person who cannot catch enough seals to feed a family, or somebody who needs many bullets to kill a seal. When used derisively, as he writes, the difference in intonation is unmistakable (Nooter 1976: 9). Thus, teasing relates to the topic of leadership, in that it is part of the negotiation of status and esteem, and may influence the decisions of travellers on whom to follow or whom not. Many times I have experienced the power and impact of teasing during my fieldwork, be it through being teased myself, or observing other people being subjected to it. Slightly related to Nooter's example of the derisive use of *piniartorssuaq*, on a number of occasions I was teased about wanting to become a 'hunter's wife' (*nuliaq piniartoq*), something which I definitely did not strive for. These jokes related to my wide interest in all kinds of tasks in daily life, which also included various issues important to hunters' wives, such as the processing of the catch. This attention was sometimes not quite understandable for some Sermilingaarmeeq (outlining my research project as an explanation was alien to many villagers). Not wanting to continue being subject to extensive teasing when inquiring how to skin and cut up a seal, for example, I stopped asking after a while.

Another example of the effects of teasing and wayfinding, which had much more serious consequences, is a story I heard whilst spending some weeks in Ittoqqortoormiit, a small settlement some hundred kilometres north on the east coast from the Ammassalik area. An employee of the weather station in town told me the following: In the 1970s another group of people from the Ammassalik area had arrived in Ittoqqortoormiit by ship in order to settle there. But the people had a hard time being accepted by the local inhabitants. When asking for good hunting grounds, the locals repeatedly gave them wrong information. As a result, they travelled whole days without finding the right locations, and were unable to catch anything. The Ittoqqortoormeerneeq found that funny, and did not hold back with their teasing. These settlers soon left the area. Thus, teasing, shaming, and derisive talk can be quite hurtful at times, and in severe cases it can result in leaving a settlement. As Petersen writes, 'ridicule was [and still is] a serious "punishment" in the Inuit society' (2003: 89, brackets added). On the other hand, teasing forms part of most conversations in East Greenland, and it also has an integrative function. If people stop bantering about someone, this can imply a (temporary) social exclusion from the community (cf. Briggs 1970).

Moreover, the above example from Ittoqqortoormiit also illustrates that information on fishing and hunting grounds is not always freely passed on to others, as was the case with communication among Sermilingaarmeeq about good fishing spots in the fjord near the village that I have mentioned above. In this and other similar cases of communication with strangers who are not regarded as proper members of the community, this information may be held back, and communication therefore affects not only individual movement and freedom, but also group cohesion and solidarity. The complex roles of different audiences and processes of inclusion and exclusion also come to the fore during communication via the inter-boat radio, which is often used to pass on information about wayfinding and orientation whilst travelling.

Communication via the inter-boat radio

The very high frequency (VHF) radio is a means of communication widely used among hunters/fishermen and other local travellers (cf. Aporta 2004: 16 on Canadian Inuit). Mobile phones, which are very popular among the inhabitants, are only sometimes relevant in this regard, as the availability of network cover limits their use during travel to few particular areas and the vicinity of the settlements. Radio transmission also does not work everywhere in the wider region travelled by Tunumeeq, but it covers large areas of the fjord system. East Greenlanders en route often have their VHF radio sets turned on in order to communicate with other travellers, and to listen to news. Most boats in the region are equipped with a radio set. According to official regulations, every boat using the radio service must hold a valid boat licence. The radio system offers a number of different channels, each with specific functions. The public channel No. 16 is always kept clear for information such as weather forecasts, important news, and emergency calls. Hunters and fishermen communicate amongst each other via channel No. 8, and for some of the hunters this is the main (and sometimes only) channel they listen to. An employee of the radio station in Tasiilaq mentioned to me that sometimes boat owners not holding a valid boat licence do not switch on channel 16 in order not to be traceable by the police. Apart from channels No. 8 and 16, two more channels can be used for conversations among individual boats that are not meant to be heard by the general public. Usually two parties call each other on one of the public channels (e.g. No. 8) in order to arrange the continuation of their conversation on one of those channels that cannot be heard by the other radios. Through these channels it is also possible for a boat to get connected to an ordinary telephone.

The radio facilitates, among other things, the exchange of information on issues related to wayfinding, such as relevant environmental conditions, weather or ice prevalence, or the navigability of travel routes. Especially for boats in trouble, such as in cases of accident or breakdown, the radio is an important means to call for help. Different audiences can be chosen according to particular circumstances and preferences, using the different available channels. People back in the settlements, such as family members, sometimes listen to the radio in order to trace the location of particular travellers, to learn about their plans, or their success in hunting or fishing. They can also send messages themselves and pass on relevant information. Nonetheless, not all households have a radio set at home, and as far as I know, radios do not serve as a means of communication among people in the settlements (who rather rely on telephone or mobile phone).

I will give an example of the villagers' use of the radio. One evening in early May 2007, Gedion was en route fishing with some friends. I was visiting at the home of my friends Julia and Peter. Julia is Gedion's aunt. The radio was running and Julia and Peter were following the communication between Gedion and some other boats. As with the trips I have described above, there was much ice in the fjords, and the chances of being able to get back to Sermiligaaq were uncertain. During that evening Julia was on the telephone a few times with Gedion's mother, her sister, who lives with her family on the other side of the village, exchanging news about the young man's trip. They were all paying close attention to the information they could get via the radio, but did not use the radio to send messages.

By and large, boat drivers make extensive use of the radio for communication amongst each other. Meetings are arranged, and information exchanged, for instance on the prevalence of fish, hunting possibilities, or on other issues such as camping locations, possibilities for getting a lift, and the like. During the boat trip with Nikolaj I described above, for example, at one point he arranged a meeting via the radio with a red cabin boat owned by a hunter from Tiilerilaaq. We got together at the edge of an ice floe, where one of the crewmembers of the red boat, a young Tiilerilaarmeeq, helped Nikolaj and his apprentice cut up a big bearded seal (*anneq*) that Nikolaj had caught. Some of the meat was shared with the other hunters. Then, some hours later, the hunters from the red boat announced via channel 8 that they were cooking meat (from that bearded seal), and invited us to join them for a lunch break and some seal meat. Thus, shortly after, we met up with them as well as two other boats with some other hunters in one of the fjords. We had seal meat on the red cabin boat, and then stayed for a while at that location socialising, relaxing, and enjoying the nice weather. The men were talking about their catch, cleaning their guns, and making lots of jokes.

Calling other people in order to ask for a favour whilst travelling (or to pass on relevant information) is a very common occurrence. I will illustrate this with a similar incident, which took place one day when I joined Bianco, a hunter from Sermiligaaq, during his hunting activities in the waters near the village. Bianco had caught a big bearded seal, which filled up more than half of his small motorboat, hindering our movement on board. The weight of the animal also slowed down the boat. Therefore, Bianco called some fishermen from Sermiligaaq via the radio. They had a bigger cabin boat and were fishing nearby. He asked them to take over the seal, and to transport it back to the village for him. A little later, with more space in the boat as well as less weight, we continued the hunting trip. Many times during various boat rides, the boat crew I had joined was called by other boats (often driven by family members or friends) that were telling us to come to a location abundant with fish. Sometimes, we then proceeded to this place, finding a variety of other boats with people busy fishing, and quickly joined in. In other cases, once arrived at the indicated locality, the fish (and the other boats) had already proceeded elsewhere (cf. Pálsson's description of the sociality of a fleet in the Icelandic fisheries, 1994: 915).

From time to time the radio is also used for entertainment and making jokes, as frequently happens among the inhabitants. Local travellers often do not mention their names when calling a boat (it is the same when they make telephone calls), since within the small population in East Greenland people often know each other by voice. I will briefly recount a situation in which I fell victim to a joke relating to my limited skills in recognising other people by voice. This joke also illustrates some general features of conversational styles common in East Greenland. It was June 2007. A group of good friends from Sermiligaaq and I were on our way back from an unsuccessful narwhal hunting trip in the Sermilik Fjord. We were driving in one of the bigger cabin boats. Vittus, the boat owner, let me take over the steering wheel for a little while. Shortly after, somebody was calling our boat via the radio: '*Elisa, Elisa!*' (the name of our boat). I was the one to take it. I asked '*qaneq?*' (what's up?), and the man replied by asking me who I was. '*Ittit kia?*' I told him my name, and then enquired of his. He did not give me his name though, but answered instead '*unarolinga*' (I am your boyfriend) – very obviously making fun of me. My companions, when I asked them who it was, burst out laughing and would not tell me the name of the person who called, saying that they did not know. I should know best, they said. They went on teasing for a while, speculating about whom from among the men of Tiilerilaaq, where we had spent time in the previous days, I had chosen as a boyfriend. Later on, with several boats having passed our way, I had figured out who the joker was: one Sermilingaarmeeq who was also en route and who had made similar jokes before. I did not bring up that joke when meeting him the next time.

This incident illustrates some characteristics of radio communication which relate to ambiguity and the complex role of the audience. Here I draw parallels with what Briggs (2000a: 121–3) has written about conflict management in an Inuit community in Canada. That joke, on the one hand, was only possible since the addresser was at a distance. It built on the contingent impossibility of face-to-face contact. Accordingly, though later on I thought to have recognised the joker, I would have never known for sure. Thus, basically I would not have been able to confront that person with it. Although the effect of that joke was based on my (the foreign anthropologist's) limited skills in recognising other people's voices, on a broader level, too, Tunumeeq do not always recognise the voices of other inhabitants, especially if these come from settlements with which they are not very familiar. Therefore, cases of radio communication among Tunu-meeq, who most usually do not mention their personal names unless asked for them, entails a degree of ambiguity, depending on the relationships of the people involved. But on the other hand, the effect of the joke was also tied to the addresser's expectation that some kind of audience (i.e. my boat companions, or other listeners) would listen to it, or would be told about it, and would pick up the joke and tease me about it. Thus, he relied – at least additionally – on an effect created by some sort of publicity.

I want to return to the observation that usually the name of a boat is called via the radio, and not the name of a specific passenger. Thus, at times it appeared as if boats communicated with each other and not passengers. Boats and passengers appeared to merge into each other, forming a new entity. Here, the boats constitute places or knots, entanglements of the people's lines of wayfaring (Ingold 2007). In relation to these places I noticed, whilst travelling with East Greenlander, that the calling party sometimes recognised the voice of the person answering radio calls, but sometimes not. In the latter case, however, this information was often not requested (contrary to the example in which I answered the radio), and this ambiguity of not knowing exactly to whom one was talking was not seen as something that necessarily had to be resolved. In a region as small as the Ammassalik district, most boats, especially the bigger ones and the vessels frequently cruising regional waters, are well known by most inhabitants. Moreover, a boat is often associated with a particular group of people who are expected to be using the radio set. These groups most generally comprise people who are known to be hanging out together in daily life, often the wider kin of the owner, and regular hunting or travel partners. Often it did not seem to be of major importance, out of those people travelling together, who would answer the radio set. The calling party had a rough idea of whom they could expect to be talking to, which would be enough to satisfy them. This also points to the important role of a person's kin group for identification purposes. For example, East Greenlandic friends sometimes did not know the name of a person I was asking about but would be able to tell me the kin group he or she belonged to. Not asking a person's name during communication with the inter-boat radio might also relate to the fact that it does not always matter which person of a boat crew one is talking to, as some information could equally well be provided by most members of such a group, or at least could be quickly sought from the other passengers.

However, when I was told to answer the radio (in the example above), the situation was different. I was not that familiar with many issues concerning our trip, and also my language skills at that time hindered smooth (or fast) communication. I suspect that the man I was talking to was quite surprised to hear my voice on the radio. Moreover, I think his joke was intended to make fun of my rather ambivalent role as an anthropologist (or student learning their lifestyle and language), which was sometimes hard to fit within the categories of people to whom East Greenlanders are accustomed. My role, on the one hand, did not correspond to that of most other foreigners, who normally would not answer the radio of a Sermiligaaq boat, nor, on the other hand, did it match with the position and the knowledge of other Sermilingaarmeeq. On this point, we may compare James Howe and Joel Sherzer's (1986) account of humour among the Kuna in Panama, and particularly their discussion of the relationship between the Kuna and the anthropologists. They show that the latter are frequently the butt of jokes (as I also found during fieldwork), which is explained by the problematic social category to which anthropologists belong. Being in a position of 'betwixt and between' (*ibid*: 689) and 'not-Kuna-but-like-Kuna' (*ibid*: 690), they are trying to be like the Kuna but undoubtedly they are never able to pass as Kuna. Trickery and humorous narratives offer one way for natives to deal with this particular ambiguity.

By and large, radios are used to send a variety of different types of messages, either through one of the public channels, via one of the channels to be heard by two boats only, or via a radio to telephone connection. Different channels are chosen according to the type of message, audience, and/or incentives for communication. I will give another example. One day in late summer 2006, I was driving with an East Greenlander from one of the villages back to Tasiilaq. I was accompanying a foreign television team who had arranged this boat transfer. Having stopped at a glacier earlier on for some hours, the boat ride took longer than expected. Therefore, the boat driver called his wife back in his village via the radio connecting to her landline. He told her that he would be going to sleep in Tasiilaq that night and would be coming home the next day. The wife hardly responded, and from her sparse words as well as her tone of voice it became clear that she was not pleased by this message. She hardly said anything, only '*bye*' as one of her few words, and then hung up soon after. The boat driver gave me a knowing look, and repeated 'she said bye'. I do not know that couple well, and could only assume, from the feeling and the nonverbal signs I thought to have sensed, that some jealousy issues might be involved here. Thus, despite, and perhaps because of, the limited possibilities for nonverbal communication that radios offer, emotionally charged issues can be exchanged without explicitly speaking about them. As I have often found, these nonverbal messages are very clear and well understood by the persons directly involved. In this, apart from the choice of the words, the length of the utterances and the conversation as such, the intonation of voice and the silent breaks are significant.

My last examples of radio communication have illustrated a number of conversational features widely shared among East Greenlanders and other Inuit. Generally in East Greenland, much verbal exchange consists in short comments (including jokes) rather than in long, elaborate monologues. Moreover, interpersonal communication often takes place in a non-confrontational manner, characterised by an avoidance of unpleasant issues. Concerning the latter, radio communication shows some particularities which are especially related to distance, the role of different audiences, and the reliance upon speech. Because of the spatial distance between communicating parties, the radio allows for the exchange of comments which would not be possible in face-to-face interaction, as for example in the case of the joke I have recounted. Comments can be more direct or antagonistic, since the distance creates some ambiguity and protection from direct, face-to-face confrontation. Yet, since some other communicational modes such as gestures or facial expressions are not possible via the radio, this medium necessitates verbal expression and supports talking. Interaction via the radio, in this respect, is, to a greater extent, based upon sounds and less on other nonverbal means of communication, which are quite important in face-to-face interaction. Nevertheless, radio messages imply the possibility of communicating more than words can carry, which was illustrated by my last example of the married couple's interaction. The choice of words, the length of silent breaks in between as well as the sound of the voice become even more important. Generally, radio communication enhances participation in community life, brings people together, and fosters bonds with and among those en route. Yet, as has been illustrated by some of my examples, communication via this medium is adapted to different audiences, depending on the channels chosen, and allows for particular strategies relating to these respective audiences.

I now want to turn to a slightly different perspective. Not only do people communicate *whilst* they are travelling, through language and various nonverbal ways and at times via media such as the inter-boat radio, but taking off and leaving a place, i.e. moving away, is also recognised as a clear sign of not wanting to communicate with particular people and of trying to avoid meeting them. Thus, moving both communicates and, at the same time, eliminates communication. It is a widespread way of dealing with confrontational situations among Inuit.

'Movements communicate': dealing with conflicts

During my fieldwork in East Greenland I have experienced various situations during which people left, for example to another village or to town, in order to avoid having to deal with possibly unpleasant situations. The non-confrontational style, which I experienced in various realms of daily life in Sermiligaaq and other East Greenlandic settlements, has been reported from many hunting and gathering societies (cf. Woodburn 1982),[9] and is often mentioned with regard to Inuit groups around the circumpolar North (cf. Bodenhorn 1993; Briggs 1970, 2000a; Larsen 1992b). In Greenland more generally, Larsen (1992b) tells us, the approach to solve conflicts verbally is not very pronounced, yet there are other ways of dealing with confrontational situations. Inge Kleivan, writing about West Greenland, sums it up:

> It was considered unbecoming to show strong feelings at normal times, but there were nevertheless various ways of demonstrating that not everything was as it ought to be. The clearest way of showing it was simply to move elsewhere. The mobility which characterised the Eskimos allowed open conflict to be avoided in the first instance by moving away.
>
> (1971: 16)

This observation has resonances with the situation in East Greenland, in bygone times as well as today. Taking into account various changes in ways of dealing with conflicts in contemporary East Greenland, it is still the case that moving is a way of communicating without words. In the context of the settled lifestyle of today, this is supported by the fact that most inhabitants have relatives or friends in other settlements with whom they can spontaneously go and live (and often do so for one reason or another). I will give some examples.

Moving can be an effective way to decline a request indirectly. For instance, one day in summer 2008 I was looking for a possibility to get to Kulusuk by boat, and it turned out that Filippus, the husband of a good friend of mine, who had offered

Figure 3.8 Sermilik Fjord, June 2007
(Photo: Sophie Elixhauser)

to drive me, was tied up with business. Subsequently, Filippus and a few other villagers were asking around for me to find another travel possibility. They found out about a young man who was planning to travel the same route. He would leave a day earlier than I had projected, but that would have been fine with me. Having heard about this possible option, I proceeded to the harbour, and asked the men at the pier about this planned boat trip. The boat driver had not yet shown up, and the men could not say if he would agree to give me a lift. I was told to wait. After a while, the men – wanting to help – suggested that I should already pack my stuff and come down to the harbour, since '*uppa ajingilaq*' (it probably might be all right). After doing so and just walking back to the pier, some Sermilingaarmeeq I met on the way told me that this young man had just left the village. The men at the harbour had most certainly informed him about my situation. His quick take off was a clear sign that he had not wanted to give me a lift. He had left before I could have met him, thereby avoiding an unpleasant confrontational situation. Afterwards, a few villagers in the vicinity, some of who had observed what had happened, were talking about this incident. A few people asked me '*qimapulit?*' (did he leave you behind?), which was followed by my lifting the eyebrows, signalling a yes. Some of them went on to ask '*sooq?*' (why?), and I answered '*nalivarnga*' (I don't know). In some other cases I was further told '*nattinaraalit*' (poor you), and given a pitying look. Through the choice of words and especially the accompanying facial expressions, I noticed that some of the Sermilingaar-meeq I had been talking to did not consider the young man's behaviour very nice. However, nobody would verbally express any criticism. As with giving orders and decisions on following a situational leader, the young man was fully entitled to decide to travel by himself and not to give any explanation for that. This right is connected to the notion of personal autonomy, which implies that it is inappropriate to try to convince anybody of another person's opinion or to judge their decisions. It was up to the young man (*nammeq*) to decide on his actions. Others might try to understand, as for instance through asking me 'why?', but would not formulate any open criticism. This has been well expressed by Briggs in relation to Canadian Inuit. She states that 'the necessity to justify one's actions is an intrusion on what would be a legitimate sphere of privacy' (Briggs 2000a: 116). I did not see that man after the incident, since I left East Greenland shortly after. But if I had met him, I would not have tried to ask him about it. As I have learned during my fieldwork, Tunumeeq do not usually interfere with one another's actions, or try to influence or inquire into each other's intentions, plans, or motives.

Relocations within the wider region of East Greenland (or sometimes West Greenland, or Denmark) are often connected to partnerships breaking up, or to jealousy. For example, in cases of divorce or couples breaking up, which I observed a number of times during my time in Sermiligaaq, one of the two usually moved to Tasiilaq or another village in the region. Apart from avoiding having to meet the ex-husband or boyfriend in daily life, this also gives better possibilities of finding a new partner as I was told on several occasions. Furthermore, relocations are often considered to be the only solution for women wanting to break out of a troublesome partnership. Concerning the latter I have heard of some cases, in Tasiilaq, of alcohol abuse and violence within marriages, in which ending a partnership was only possible through moving away and getting out of reach of the violent partner.

I want to elaborate upon one of these examples, which I heard about during my stay in Ittoqqortoormiit in Northeast Greenland. The details have been recorded in a recent collection of life histories (Christensen and Bang 2007: 11–29).[10] A woman from Ittoqqortoormiit used to live in Tasiilaq. There she was married to Salomon, originally from Tasiilaq, and they had two children. Yet Salomon, time and again, became violent towards his wife, due to his high intake of alcohol. When these incidents began to happen on a regular basis and Salomon also began to abuse the children, Therecie knew that she had to get away from him. But she also knew that if she would tell her intentions to her husband, her life would be in danger. Salomon would never allow her and the children to leave him, and would react aggressively.[11] The only possibility Therecie saw was to move back to Ittoqqortoormiit, to the place where she had grown up and where she had family and friends. There she would be out of reach of Salomon (airplane tickets are expensive). She started to organise her escape to Ittoqqortoormiit. This attempt was not easy since in a small town like Tasiilaq, news of travel plans quickly spreads, and it had to remain an absolute secret that she and her children were booked on the flight to Ittoqqortoormiit. Yet she was supported by the help of good friends, and also cooperated with the police. The escape was successful, and Therecie started a new life back in the place where she had grown up.

This example is one of a number of similar cases in which people leave a place in order to eliminate communication. However, this avoidance can also be interpreted in a slightly different manner, as in some cases it might also be a form of respect, in the sense that the avoided person is never directly repulsed and can thus keep face in the community. Furthermore, since from my experience, the reasons why a person moves away are often not explicitly stated to others, the other villagers get less of an incentive to gossip about interpersonal tensions and disagreements.

Conclusion

Communicative encounters during processes of travel are influenced and created through a variety of features which, above all, involve the people communicating, who use the contexts and means of travel in specific ways, in order to establish connections with other human and non-human beings, or at times also to disconnect them or to avoid their formation. In the same way, technological and environmental conditions influence people's communicative encounters and the places in which they occur. On the one hand, different kinds and contexts of travel influence and give priority to specific and varying modes of communication. For example, interactions during boat travel include numerous gestures and bodily movements, influenced among other things by factors that impede hearing such as the noise of the boat motor or the distance involved when boats meet on the water. Radio communication, too, is tightly aligned to the particular setting of travel. The features of this technology give priority to speech, but do not rule out the ways of communicating without words that are so widespread among the Iivit. On the other hand, the examples given have illustrated conversational patterns that also appear in other settings of daily life (such as in the settlements or at home), in contexts of learning, leadership and giving orders, and the avoidance of confrontational situations. All the examples highlight the value placed on personal autonomy; they also reveal the power of focal vision, of mimicry, and of speaking about something directly, which may be circumvented by various forms of nonverbal and indirect forms of verbal communication. Hence, not only when travelling, people often communicate whilst moving along side-by-side, without directly looking into the other person's face (or only briefly); they correspond through peripheral areas of the visual field, through brief comments, gestures, and silences. With regard to speech, too, a direct approach is often avoided by means of particular ways of speaking that entail ambiguities of various kinds, such as through the medium of the VHF radio or through humour.

All in all, I believe the indirection that I observed in many communicational practices is not only related to the high value placed on personal autonomy, but also tells us something more fundamental about the notion of the person in East Greenland. A person, whether human or non-human, seems to be particularly vulnerable and needs to be protected in one way or another. Accordingly, people are careful in how they approach others, avoiding face-to-face encounters by moving side-by-side, allowing gestures and silences to speak for themselves, and not telling others what to do while leaving them to find out for themselves. The spoken word seems particularly powerful, and various forms of verbal indirection are used to circumvent the threat it poses.

The themes of movement and communication offer a variety of additional perspectives which I have not been able to deal with. One of these is looking at how inhabitants move within specific buildings such as the residential house, a theme to which we turn now.

Notes

1 'Landscape' and 'environment' are used interchangeably in this book.
2 For an analysis of East Greenlandic animal names see Dorais (1984). Dorais explains that on a general level these names, as well as other parts of the East Greenlandic vocabulary, imply an avoidance of confrontation with dangerous environmental forces.
3 Similarly, ethno linguist Michèle Therrien mentions an animal language in 'primordial' times, resembling an Inuit dialect. This category of animals, she explains, could also comprise human beings (1987: 113).
4 Speaking about respect, I do not want to convey a general image of the Inuit as 'ecologically noble savages', who have always 'lived in unity with nature', as they are sometimes being portrayed (or portraying themselves) outside of and also within Greenland. From my experience, the East Greenlandic Inuit may show deep respect towards certain beings and elements of the non-human environment whilst at the same time acting environmentally destructive in other realms. On the contradictory perception of Inuit as guardians and destructors of nature, see Sowa (2014).
5 Pure hunting trips are most often only joined by men, yet, from time to time, family fishing trips and weekend excursions also include hunting activities, depending on the mood of the male passengers and whether animals are spotted. Female hunters are rare. One female friend, who is married to a professional hunter and sometimes likes to join

him on his hunting trips, told me that she had taken a seal before. Yet, during my time in East Greenland, I never experienced a woman hunting.

6 Nowadays there are also routes for snowmobile travel in the winter time, but these are mainly limited to areas around Tasiilaq and Tiniteqilaq.

7 For example, during the initial months of my fieldwork I once went on a fishing trip with my host family. Shortly after disembarking I asked my friend Maline about the destination of our trip. She told me that she did not know those sorts of things, and that I should ask her husband, Lars, who was driving the boat.

8 Henriksen reports very similarly of the Naskapi in Canada, who 'do not tolerate any meddling from others in their decision-making', and place a high value on personal autonomy (1993: 44). This leads to difficulties in arriving at mutual decisions, as is also sometimes the case in East Greenland.

9 Kirk Endicott, for example, describes movement to avoid potential or real conflict among the Batek in Malaysia (Endicott 1998: 122). I have encountered similar situations among the Agta in the Philippines (Elixhauser 2006: 77).

10 I am using the pseudonyms from Christensen and Bang's book (2007).

11 For a detailed study of men's violence against women in Greenland, see Wagner Sørensen (1994, 2001).

4   Family life

The power of words, personal space, and the materiality of a house

Houses and architecture more generally have attracted little attention among anthropologists (Carsten and Hugh-Jones 1995; Humphrey 1988). Where they are considered, it has often been in contexts of symbolism and cosmology rather than as objects of inquiry in themselves (but see Anderson et al. 2013). For example, in Claude Lévi-Strauss's notion of 'house societies' (sociétés à maison), the house is regarded as a key principle of social organisation (1987: 151). Houses are seen as units of society that are hierarchically ranked. This approach is difficult to apply to non-hierarchical societies such as East Greenland. Moreover, Lévi-Strauss's structural approach does not consider the architectural features of houses, and how the inhabitants experience them (Carsten and Hugh-Jones 1995: 12). Also, Pierre Bourdieu's study on the Kabyle house (1979) is written from a rather structural point of view, though later in his career Bourdieu critiqued his own work (e.g., through his concept of habitus) and became one of the leading 'poststructuralists' (Goodman and Silverstein 2009).

This chapter explores everyday communication and relatedness among family members and people living together. Most of my examples will concern people's interactions in their homes, as encounters in this setting are often linked to a certain kind of familiarity. Janet Carsten writes: 'The house brings together spatial representations, everyday living, meals, cooking and the sharing of resources with the often intimate relations of those who inhabit this shared space' (2004: 35). With respect to the examples drawn from the setting of the house, it is important to note that I do not consider the house merely as a material backdrop for the inhabitants' communicative practices. Contrary to some social constructionist studies, I direct attention towards the materiality of the house and its spatiality as experienced by the inhabitants (cf. Carsten and Hugh-Jones 1995; Carsten 2004: ch. 2; Turnbull 2002). The house structures the inhabitants' movements and communicative encounters by its rooms, doors, and windows. As David Turnbull writes, 'buildings perform people by constraining their movements and by making likely certain kinds of encounters between them and others' (2002: 135). Moreover, houses not only relate to inhabitants' present day activities, but also have a history and enfold memories of the past. How life was conducted in the houses people inhabited in the past influences present day lifestyles and social interactions (Carsten 2004: 31–56). As Edward Casey (1996: 39) argues, drawing upon the phenomenologists Gaston Bachelard and Martin Heidegger, 'it is in dwellings that we are most acutely sensitive to the effects of places upon our lives. Their "intimate immensity" allows them to condense the duration and historicity of inhabitation in one architecturally structured place'.

Life in turf and stone houses

When the Ammassalik region was first documented at the end of the 19th century, East Greenlanders lived in long multi-family houses during the winter, which were built from sods, stones, and timber (Holm 1911: 35–8). The walls were made of stones, with grass-turf in between, and the inside of the walls was partially covered by skins. The beams in the house, the platform, and house-props were made from driftwood. Small windows with gut-skin panes looked out towards the sea (Thalbitzer 1914b: 354). Occupancy frequently changed, and at the end of a summer groups of people were allowed to move into any available empty house. These communal houses were inhabited up until the 1920s. Subsequently, winter settlements became more permanent, while the average number of occupants per house declined.[1] Turf houses (*ittilorat*) became smaller and changed to single-family homes (see Figure 4.1). Apart from size, these smaller turf homes were initially built and set up in a similar manner to the bigger communal houses. Factors contributing to this development included an increase in personal property, which necessitated that people returned to live at the same house each winter. Petersen (2003: 117) names the increasing use of (imported) wood in houses as a crucial factor in this respect. Wood was owned individually by fathers of families, which was linked to the fact that skins sold were the property of individual families (since wood was purchased with the income). With less turnover in the settlements' inhabitants, the number of very small settlements decreased, since isolation for several years was not found very attractive. After 1950, the one-family turf houses were gradually replaced by the imported Scandinavian buildings inhabited by East Greenlanders today (Petersen 1984a, 1984b).

The communal houses were around seven to fifteen meters long and four to five meters high. They accommodated an average of four to six families, with cases known of up to eleven families. Each settlement had one such communal house, with all inhabitants living in one room (Holm 1911: Fig. 31; Lee et al. 2003: 11, 171–2; Thalbitzer 1914a: 352–64). Life went on in a very confined space. The large, rectangular or oblong buildings were free standing or built with the back dug into a hillside. The entrance in the front wall was through a low passageway. The back part of the house was taken up by a wooden sleeping platform which extended for the entire length of the building. It was divided into partitions by stretched seal skins fastened to a number of poles. Each partition was home to a nuclear family consisting of the married people, their unmarried daughters, and small children. Unmarried men, older boys, and guests slept on a smaller version of the platform on the side or front walls of the house (Holm 1911: 37–8). Holm tells us that the women spent the greater time of the day sitting on the platform with their legs crossed preparing skins, or occupied with sewing, handicrafts, cooking, and looking after the children. If the men were not working on their hunting utensils, they often did not do much more whilst in the house than eat, sleep, tell stories about their hunting adventures, and practise drum-singing (*ibid*: 60).

The entrance tunnel (*sorsooq*) was where the rubbish was kept, including parts of caught animals and so forth. The washing also took place here. Greenlandic houses were kept very warm, and the preservation of heat was supported by the shape of the houses, the insulated walls, and the number of people living together. The residents often wore no more than the barest underclothes indoors, and regularly sat on the floor (*naleq*) where it was colder (Lee et al. 2003: 19; Holm 1911: 60). The winter dwellings were occupied from August or September to April or May. In the summer time, when East Greenlanders moved into their tents, the roofs of the houses were taken off, as skins and wood were needed for the tents. The rain had cleansed the buildings by the time the people moved back in the autumn (Petersen 2003: 58). Property existed only in relation to very few items such as some personal things. Hunting grounds and also the walls of the turf houses were communally used.

Generally, women spent more time of the day in the house than the men and were responsible for more of what took place inside (e.g., housework, education of children). Accordingly, many arctic scholars regard the house among Inuit as a female domain. An explicit link is drawn between women, houses, and wombs (e.g., Oosten 1986; Saladin d'Anglure 1986; Therrien 1987; on East Greenland Robert-Lamblin 1981). Barbara Bodenhorn (1993), however, argues against this analogy in the case of the Inupiat in Alaska, referring to discussions initiated by Rosaldo (1974). She states that Inupiaq houses were divided into male *and* female spaces. This view has been supported by the archaeologist Peter Whitridge (2004), amongst others, who describes the distribution of women's and men's places in prehistoric Inuit houses. It is consistent with the available material from East Greenland.

East Greenlandic long houses accommodated several extended families, with each nuclear family inhabiting one partition of the communal platform. Nuclear families from the same extended family had their partitions grouped together. Crossovers between a nuclear and an extended family were fluid, and a nuclear family consisting of father, mother, and children plus any foster children would easily develop into an extended family, or the latter could split up into several nuclear families (Petersen 2003: 90). Residence was most often virilocal with the daughter moving in with her husband's family, but alternate stays with the parents of both couples are also reported. Widows went to live with their closest family members. If their nuclear family was not available, they were often accepted into a cousin's household (*ibid*: 91).

A collectivity of families in a communal house formed a loose arrangement without formal organisation. Food was commonly exchanged between housemates, and entertainment and other forms of sociality were shared. Apart from that, individual households retained their independence. During summer times, when East Greenlanders moved into tents, they split up into smaller family groups. Only sometimes did a family group return to live at the same location in the autumn, and one could observe a regular turnover of co-settlers (Petersen

1984b: 625, 2003: 70–84). Families sharing a communal house were usually related, and East Greenlanders only occasionally lived with people who were not close relatives. Petersen (2003) suggests that this tells us something about the communal house as a unit. Since the people were forced to live together in one room for seven to eight months of the year, this could easily lead to tensions, especially with people who were unrelated. Yet raising a quarrel was not appropriate, and mechanisms for dealing with conflicts such as drum and song duels did not take place between people from the same house (on drum and song duels, see Chapter Six). Among relatives, especially among cousins, joking relationships helped to defuse interpersonal tensions. Therefore, according to Petersen, a 'tense situation probably mainly arose among people [in the house] who were not close kin, for in this situation there were no mechanisms for defusing such a situation' (*ibid*: 80, brackets added).

A few of the wooden turf-lined one-family houses can still be found today. Some are used for storage, but none is inhabited. In the majority of cases you encounter remnants made up of some parts of the stone foundation walls and the entrance tunnel. These ruins (*ittigut*) are scattered all along the fjords in the region, and only some are located in and around contemporary settlements. I have met many elderly local people who grew up in a turf and earth house, and once in a while East Greenlandic friends have pointed me to house ruins where they, their (deceased) relatives, or other families they know used to live. Comments about life in turf houses were always very positive, and these houses continue to figure in the collective imagination.

Figure 4.1 Turf house reconstructed by Sermiligaaq pupils, Sermiligaaq Fjord, July 2017
(Photo: Martl Jung)

Contemporary houses: characteristics and inhabitation

Nowadays, Sermilingaarmeeq and most other inhabitants of East Greenland live in wooden houses (*ittit*, sg. *itteq*), for which construction sets are imported from Denmark. A number of different types of houses are common throughout the region, and the residences in Sermiligaaq represent a handful of these types. Apart from variations in exterior colour and furnishings, many of the buildings look very much alike.

The majority of the houses are not private property in the sense that official title lies with the tenants. The people often put up the buildings themselves with ready-made materials provided by the municipality at little cost (in Danish *selvbyggerhus*, 'self-build-house'). For a substantial number of years, tenants then have to pay an annual lease to the municipality before they take over full owner-ship. If they move out before that period ends, or if the lease has not been paid on a regular basis, they lose this entitlement. For this reason, the municipality still officially holds many homes. In addition, a number of buildings are owned by one of the local companies or institutions, the municipality, or the Greenland Housing Association INI, and are rented out for a monthly lease (Morten Steen, Ammas-salik municipality, personal communication, May 14, 2007).

In general, identification with any particular building is not very pronounced among the Tunumeeq, a fact that might relate to their former semi-nomadic lifestyle. During my time in East Greenland it happened rather frequently that a household (or part of it that would form a new household) would move to another building which was perceived to be more advantageous. Decisions about moving were not taken in advance and little was needed by way of preparation. Moving house is a topic that is not greatly talked about in the community, and I have never heard people mention that they miss a particular building they had occupied for some period of time.

Apart from four modern semi-detached houses, all buildings in Sermiligaaq are single-family homes that have either one or two storeys (e.g. see Figure 4.2).[2] Many Sermiligaaq households comprise a substantial number of people and, as in former times, it is not uncommon to find three generations living together (cf. Buijs 2002). The houses are usually inhabited by a nuclear family, sometimes supplemented by partners/spouses of the children's generation and their children as well as a varying number of individuals from the wider extended family. These may be temporary or long-term household members. Yet the number of occupants is often less evident in the daytime as children spend a lot of time outside, even in the winter, and other household members, especially the men, are outdoors a lot as well. Only few households in Sermiligaaq have only one resident, most often an elderly person. In the more urban Tasiilaq, the proportion of households with few members is higher.

Generally, household membership is very fluid as the number and composi-tion of people forming a household together frequently change. For example, a person might decide to stay with other relatives in the village for a while, or with kin in another village, often without planning the length of his or her stay, or

Figure 4.2 Sermiligaaq, December 2006
(Photo: Sophie Elixhauser)

every so often changing plans. Decisions on moving to another settlement might relate to a new job or a new boy- or girlfriend, to being attracted to life in town, or wanting to leave the settlement for other reasons. East Greenlanders from other settlements are usually accommodated by their closest relatives. For the most part, the residents of East Greenlandic houses are relatives, but beyond that a household is very difficult to pin down. Usually the (married) couple forms the core in a house, which is associated with the family group around it. Children are integral to most households, though they easily move between them and the number of children in any Sermiligaaq household changes frequently. A child may opt to sleep somewhere else, to visit for a while, to go to live with the grandparents, be fostered or adopted elsewhere, or start to attend boarding school or another educational institution (cf. Bodenhorn 1993, for a similar account on Inupiat households).

After getting married – and sometimes before that – young couples often apply for their own house. The municipality allocates houses according to availability. Demand is usually very high, and young families therefore often continue to live in the household of one of their parents for some time. I did not observe a general rule of either virilocal or uxorilocal residence.

In spite of many differences in architecture and internal layout, when looking at how the inhabitants arrange interior furnishing and make use of the different rooms, life in the modern East Greenlandic houses is still reminiscent of the turf houses from before, both the large multi-family dwellings and the single-family turf houses inhabited later on. Lee et al. have remarked:

Today, in the suburban-type housing imported into Eskimo communities in Greenland, Canada and Alaska, anyone familiar with spatial arrangement in underground winter houses of the past cannot fail to be struck by the similarities in how equipment like hunting gear is stowed in the newer houses. Today, the parkas come from clothiers Eddie Bauer or L.L. Bean, but the way they are hung in the entranceway replicated that of the old fur parkas. Without doubt Eskimo perceptions of space continue to be influenced by the patterns of earlier times.

(2003: 164)

I will now take a closer look at the interior of a Sermiligaaq house and how it is used by the inhabitants. Since many of the houses in the village have a similar internal layout, I will take a widespread type of house as a showcase whilst mentioning different other layouts to be encountered.

Use of space

Many East Greenlandic houses are entered through a small porch and a second entrance room; the latter leads directly into the living room or sometimes into the kitchen. The living room features a big sofa area in one corner (*nalaasarpik*) with a coffee table in front of it, a cabinet including television and radio, and some additional furniture such as shelves. The basic pieces of furniture are quite similar in most houses as they come together with the construction set for the house. Most walls, especially in the living room, are covered with a variety of framed pictures depicting family members, deceased relatives, friends, and different kinds of religious motifs, as well as certificates/awards. Similarly, cups from dog sled competitions and other items showing some sort of achievement are often displayed on the cabinet or shelves. In another corner of the living room there is often a dining table (*neqqivik*) with chairs (*itsiavik*, sg.), sometimes alternatively located in the kitchen. The kitchen (*ingarpik*) adjoins the living room and includes at least the basic equipment of a modern kitchen. Newer types of houses often have a fitted kitchen, and some of the smaller variants feature a kitchen compartment integrated into the living room. The bedrooms, usually two in number (sometimes one), are either located on the second floor, reached by the stairs from the living room or the kitchen, or, in the case of the bungalow houses, next to the living room on the ground floor.

The porch (*paaq*) is often filled with a variety of things like bags with garbage (the garbage bags are picked up by the community staff twice a week), working utensils, animal bodies and country food left behind by the hunters when entering the house, dirty shoes, children's sleds, a tray with blubber, and so on. It is not unusual that people who pass through the porch are forced to step over some bodies of seal waiting to be cut up, birds, bags with fish, and the like. This room is used in a similar fashion to the entrance tunnels of the turf houses. The porch is often used more intensely by the men, as was the entrance tunnel.

At the second entrance room everyone removes shoes and coat before entering the main living area. The bathroom door often goes off from here (yet in some older houses the toilet door is located in the kitchen). Sometimes the floor of this room is used for cutting up seals or other animals (most often done by the women), which may alternatively take place in the kitchen or outside (or in the workspace of the service house). The newer buildings in Sermiligaaq, such as the bungalow houses inhabited both by younger and elderly inhabitants, usually no longer feature an entrance area divided into two rooms. This affects people's storage habits as some of the items formerly stored in the first entrance rooms are moved to different locations in- and outside the house.

The living room forms the centre of daily family life. The sitting area with the sofa, easy chairs, and coffee table is frequented during the day and evening by both household members and visitors. This room is accessible to everyone entering the house, and I have never encountered people closing a living room door to keep other people out, which they sometimes did, however, with respect to bedrooms or the kitchen. During evenings and weekends, one often finds family groups gathering here (sometimes including friends and other villagers), in order to watch TV or DVD together; to exchange news, gossip, and various other issues; to tell or listen to stories; or just to enjoy being together with family members. Moreover, the sitting area is a popular playing space for the children, and the atmosphere in the living room is often very lively. Housewives and older women usually spend much time in the day on the living room sofa doing handicrafts, taking care of the smaller children, talking on the telephone, watching TV, and so forth. Still today, men's tasks take place to a greater extent outside the house, and you often see groups of men at some popular outside meeting places, such as near the harbour. If men are inside you frequently find them on the sofa relaxing, watching TV, or resting.[3] These observations on gender – with differences from household to household depending on the employment of the occupants – resonate with Holm's description of house life at the end of the 19th century, summarised above. In addition and more broadly, the space on and around the sofa parallels the daytime functions of the sleeping platform in the turf houses (in the night taken over by the bedrooms), which formed the centre of family life.

Temperatures in East Greenlandic houses are usually very warm, as is reported of turf houses in the past. Throughout the year, many Iivit habitually wear t-shirts and sometimes only underwear whilst inside. The most common source of heating is a petrol oven in the living room, which is sometimes also used for boiling water as some stoves feature a hotplate. Petrol is purchased at the local KNI store. Keeping the stove going and purchasing and refilling the oil is the responsibility of the men. Children and adults alike regularly sit on the floor (*naleq*), as for example when there is not enough space available on the couch or the dining table, when wanting to stretch out in order to watch TV or sleep, or when they prefer to stay somewhat separate from other inhabitants for some reason. The floor is used for a number of daily tasks, especially by the women (as was also the case in former times), such as

Figure 4.3 Crafting national dress, March 2007
(Photo: Sophie Elixhauser)

for cutting up seals, fish, and other sea mammals, or the fabrication of traditional costumes by elderly women (see Figure 4.3, cf. Buijs 2004). Also, eating sometimes takes place on the floor – particularly among elderly inhabitants – a reminder that tables and chairs have been introduced rather recently to East Greenland.

The kitchen is by and large the domain of the women. Men are hardly concerned with preparing meals and cleaning up afterwards, and spend comparably little time in this room. For example, Lars, the husband in my host family, like most men in other households I came to know, rarely contributed to the cooking. Accordingly, the kitchen provides an undisturbed place for women's chatting whilst communally preparing meals or doing the dishes, including the daughters who are often expected to help (especially with the dishes). These talks may be among women from the same generation or different generations, such as mother and daughter. Maline and I frequently had personal conversations during kitchen work, exchanging issues we would not have talked about in the living room with other people around. This was facilitated by the door that was usually closed whilst cooking so that steam and smell from pots and pans would not spread into the living room. Moreover, the kitchen is also the place where family members sometimes take a wash in the washtub (as many of the residential houses in the villages do not have running water), and as I experienced on a number of occasions, closing the door meant that the person did not want to be disturbed. Others respected this. Thus, the kitchen may afford some kind of privacy, which supports the emergence of specific topics of conversation (e.g., intimate talk among women, such as about relationships, emotions, or the like), or its use for activities which people do not want other inhabitants or visitors to watch, such as tasks related to personal hygiene. In this context, the existence of a separate room with a door affects interpersonal contact, as do a variety of other aspects of domestic life such as gender-specific divisions of labour.

With regard to privacy, it is important to note that the boundaries and contents of what is considered private differ among societies and individuals (Haviland and Haviland 1983; Moore 2003; Schoeman 1984; Vom Bruck 1997). Here, I use the term privacy to denote practices that spatially separate oneself from other people in certain situations. In an East Greenlandic context, however, I found that the desire for this kind of spatial privacy is not very strongly developed and that a withdrawal from other people does not necessarily require spatial separation, as I will explain further below. What Anja Nicole Stuckenberger says about this kind of privacy among Inuit in Nunavut also holds true for East Greenland:

> Most houses are crowded and rooms have a variety of functions rather than the functions designated to them formally. While there is little privacy in Inuit homes, it is usually quiet and people take care not to disturb each other in their occupations. (Thus, rules of conduct shape how a space is used rather than the rules associated with the functions that a room was meant to fulfil.)
>
> (2009: 6)

Most probably, the same was true in the turf homes, where people must also have had to resort to strategies of withdrawal that did not entail spatial separation.

Further possibilities to separate spatially from other people are offered by the bedrooms (*innartarpit*). They provide retreats of a sort, and to draw back to the bedroom in order to sleep is accepted behaviour, regardless of the time of day. It is not uncommon or found strange for people to sleep during the day, or similarly to stay up during night-time (cf. Gessain 1970: 109). Hence, sleeping does not always follow a regular rhythm, and sleeping during the day is not regarded as laziness.[4] When somebody retreated to the bedroom, closing the door was understood as a sign of not wanting to be disturbed. Being tired, likewise, was an accepted explanation for leaving other people's company. For instance, when visiting some of my good friends' houses in Sermiligaaq, my friends often did not want me to leave their company in order to go elsewhere. Apart from other rather obvious obligations, such as having promised to help a friend or having to go shopping, to cook, or to take care of somebody's children, being tired and wanting to sleep were among the few reasons that my friends would accept without being offended. Explanations that were related to activities which did not seem absolutely necessary, especially those not including other people (e.g. reading), were often not regarded as reasonable. This, I assume, relates on the one hand to the fact that most people I was spending time with generally preferred activities that included the company of others. On the other hand, it was known that I did not have a fixed daily rhythm and could decide upon my daily schedule more freely than most. For this reason, I could not leave the houses of good friends without explanation or excuse.

Generally, as noted before, there is a tendency for fewer and fewer people to share a house and the Iivit are getting used to having more space available. Sermiligaaq residents also sometimes express the wish for more space. I have sometimes heard members from households with many people complaining about the number of people in their homes, saying it was very tiring (*qassinara*), and in some cases giving this as a reason for frequently visiting friends' or relatives' homes with fewer occupants. Alternatively, however, most Iivit I met do not enjoy living by themselves or with only one other person. In cases where the spouse or partner was away for some time or the children had recently left to boarding school, I regularly encountered Sermilingaarmeeq asking relatives or friends (myself included) to keep them company at their home, especially during the nights. Spending the night by oneself in a room or house is felt to be frightening and sometimes several people prefer sleeping all together in one room instead of using an available free bedroom. This is also reported by Therrien (1987: 44) and Stuckenberger (2009) with respect to Inuit in Canada.

Building on this information on the internal spaces of East Greenlandic houses and their use by inhabitants, I will now concentrate on some expectations about social interactions and particular modes and situations of communication that came to my attention whilst living and visiting in East Greenlandic homes.

Television and the body: contexts for comments and teasing

As noted before, many conversations in East Greenland comprise rather short utterances and expressions, which are aligned to the immediate context of the involved persons' surroundings. One such context is the television (*piarnsiini*), which can be found in most households in Sermiligaaq (see Figure 4.4.; nowadays many households have large flatscreen TVs). The national channel broadcasts from 4 pm on weekdays and all day long on weekends, and in many households the television is turned on for most of the time during broadcasting hours.[5]

When watching television together with East Greenlandic friends, I noticed that my companions often attended mainly to single sequences and pictures. Breaks in the middle of a programme are very common, and my friends often did not seem to care for the wider context and course of a film or programme (this has also been mentioned by Therrien [1987: 45] with respect to Canadian Inuit). Moreover, sequences from TV or DVD/video programmes are frequently commented upon and formed into jokes, relating them to people or happenings from the neighbourhood. Comments on images from TV often consisted in expressions of excitement and being fond of something or, on the contrary, in showing dislike. People would often comment on favourite images with colourful flowers and trees or selections of fruit, both of which are rather rare on the Greenlandic east coast and found attractive. Negative comments regularly arose

Figure 4.4 Sermiligaaq living room, December 2006
(Photo: Sophie Elixhauser)

when insects or animals such as spiders were displayed, as many Iivit find these animals disgusting.

Pictures and scenes from TV often form the starting point for personal stories on a related topic. For example, I was once watching a TV show about the planet Mars at my friend Hansigne's house in Sermiligaaq. This started off a conversation about the fact that humans do not inhabit Mars and other planets. Hansigne commented that some sort of non-human beings might well inhabit them. She went on to tell me about her experiences of ghosts and other non-human beings in daily life. Equally, TV programmes regularly serve as a catalyst for all kinds of jokes and communal laughter involving words, sounds, grimaces, gestures, and postures. For example, KNR, the Greenlandic TV channel, frequently broadcasts animal documentaries featuring animals from all over the world. Sermilingaarmeeq are very fond of these documentaries and enjoy laughing about the strange animals' activities and appearance. In addition, animals from these programmes are regularly compared to people from the village. Particularly exotic animals are jokingly addressed with the names of other inhabitants, individuals who are sometimes present amongst the people watching the documentary and sometimes not. I, the foreign anthropologist, was also included in this practice. For instance, one day when watching a TV programme featuring animals from Africa at my friend Sibora's house, the elderly lady started to ascribe to me the flattering appellation *'apakka'* (monkey), which she accompanied with arm movements typical of monkeys' climbing activities. She related the perceived similarities to my comparatively long arms and legs that she and other East Greenlanders regarded as extraordinary (though I kept insisting that these are only striking in comparison to the Greenlandic physique). Similarly, a German friend, who was working in Tasiilaq during the summer months of 2006 and who came to visit for some days in Sermiligaaq, acquired the nickname 'frog'. Sibora would then repeat these stories once in a while, accompanied by a little spectacle in which she recounted and demonstrated with gestures, body movements, and sounds, how the monkey in the programme reminded her of me or the frog of my friend. The frog performance was now and then supported by Sibora's nine-year-old granddaughter (and foster daughter) Karline, jumping around the room in a frog-like manner. Many times in the following months these jokes entertained members of the wider family and other villagers who came to visit.

I have experienced many other examples of Sermilingaarmeeq comparing villagers – or also themselves – with animals, funny looking persons, or various kinds of images from TV or DVDs. For instance, we were once watching a videotape of traditional drum dance performances from the Ammassalik region, and a female performer with a prominent chin was addressed as 'monkey' by my friend Julia. Similarly, though not in relation to TV programmes, I was told about the wife of a teacher with a nose 'like an elephant', which was demonstrated to me very skilfully through arm movements imitating an elephant's nose. Practices of comparing people to animals seem to be quite common throughout Greenland; for instance, I have heard these kinds of jokes when visiting the Greenlandic capital Nuuk.

In general, these examples underline the great importance of verbal and nonverbal playfulness in East Greenland, as teasing can be regarded as a type of social play which serves as an entertainer and time filler (cf. Bateson 1972). This playfulness absorbs the whole body: body parts (as well as other available objects, and even sometimes animals) are regularly used as 'tools' to visualise the joke being performed. Speaking of Canadian Inuit, Briggs has described the 'ability to conceive of, and to use, the body as a rich collection of multipurpose tools' (1991: 275), and my experiences in East Greenland corroborate this. On numerous occasions people would measure the length of my upper leg with their arms and compare it to their much shorter limbs. This was always an incentive for much laughter.

Thus, as with animal nicknames, I frequently observed, and was myself part of, jokes and laughing that involved physical issues or bodily functions. These could involve bantering about someone's big nose, bodily figures or posture, funny movements, or the wider theme of sexuality, the sexual act, and body parts and movements. In addition, laughter was often induced by somebody puking, punching a person in the butt, and so forth. This attention to bodily appearances and movements was also apparent when watching TV or DVD with friends and informants. For example, when watching the footage from Anni's and my filming activities with the Sermilingaarmeeq, or later on the final version of the documentary, the sequences that drew most attention were related to other villagers and their outer appearances and movements. One shot in the documentary that regularly caused laughter among the spectators shows Maline's son Gerth performing an interactive computer game, jumping back and forth in front of the TV screen and moving in a very funny way. I found that Tunumeeq are not easily insulted by jokes concerning their own physical features or appearance. In effect, these joke performances amount to a way of perceiving and of dealing with one's own body. The body and the physical issues connected with it are relevant here in several ways: First, communication is embodied and the body forms the basis of any communication. Second, the body and its parts are at times objectified and used as 'illustrative material', such as when demonstrating an elephant nose. Third, the body and its appearances and functions may serve as a topic or incentive for interpersonal communication.

On the other hand, particular bodily issues such as uncontrolled bodily noises are often ignored and are neither accompanied by any apology from the originator nor remarked upon by surrounding people (and they are joked about seldom). Mary Douglas has stated that social factors 'govern the thresholds of tolerance of bodily relaxation and control' (1975: 75). I argue that in East Greenland these thresholds are relatively low as compared to many other Euro-American countries. Accordingly, one day, friends of mine in Sermiligaaq had been asking me why bodily noises are so much less common among foreigners. They were quite surprised when I explained that people in some other countries are taught to suppress these noises.

Nonetheless, jokes in East Greenland, more broadly, may carry implicit meanings that are meant to embarrass the target and to be taken personally. Humour provides an important means to express dissonance and criticism, brought forward through hints and irony. Teasing thus also expresses expectations of 'right' conduct, as I will elaborate upon later in this book.

Let me now return to my initial example of animal nicknames. It struck me that animal names used for a specific person and other similar jokes or statements were hardly picked up subsequently by other villagers. For example, I was only rarely called *apakka* by anyone else except Sibora. Whenever such comparison was mentioned by another person, he or she usually cited the individual who had coined it. The addressee, then, was given the possibility to communicate his or her like or dislike, and sometimes was even explicitly asked for her opinion. This observation illustrates a general carefulness in taking over a way of speaking introduced by someone else, which relates to the power of spoken words in East Greenland (cf. Holm 1911: 88). It also reveals the value of personal autonomy, which 'precludes doing stories without permission or cooperation from the subject' and 'also militates against embarrassing individuals', as Kate Madden points out with respect to the Canadian Inuit (1997: 304–5). However, although it is not appropriate to embarrass or criticise a person explicitly, criticism and other kinds of verbal attack may well take place via the medium of humour or different forms of gossip.

Generally, throughout my fieldwork I have learned much through Greenlanders' commenting on materials such as film footage, different types of TV programmes, and photos. I found that Tunumeeq are often more open taking up topics whilst sitting around the TV, whereas too much questioning (on the part of the anthropologist or also children), whilst engaging in practical activities that require comparatively intense concentration, is not appreciated. When my friend Anni and I were working on our documentary film, for instance, showing footage and asking questions (and paying attention to comments and jokes) was a particularly valuable research tool. But when trying to learn to knit socks from my good old friend Sibora, she was not prepared to give me verbal instructions, as she was busy finishing her own woollen sock. In contrast, sitting with Greenlandic friends and informants on the sofa, relaxing and watching TV, provided a perfect situation in which people were willing to explain things to me and did not mind questions, especially if these related to the TV programme or, likewise, to interior parts or accessories of the house. It is common for visitors to ask about the decorations of walls or furniture, such as pictures, awards, certificates, and so forth, and hosts are usually willing to explain their meanings and backgrounds. Sometimes these questions serve as an incentive for longer stories about the people depicted.

Individual knowledge, reported speech, and the power of words

All in all, I found that most of the time the person telling a story or articulating a remark was speaking from first-hand experience. If relying on another person's account, the source of information clearly had to be stated, as I have noticed in many different contexts (see the example of animal nicknames). Pedersen (2009b) reports very similar kinds of storytelling in West Greenland. Referring to the genre of horror stories, she writes: 'The narrator is most often a first person narrator, who frequently makes a third person tell his or her version in first person word-by-word – again to make sure the story transmitted is as close to the truth as possible – the value of a Greenlandic horror story is generally measured by its degree of authenticity' (*ibid*: 8). Also with respect to stories that have been told by deceased East Greenlanders, the sources of information are usually stated (cf. Bodenhorn 1997: 126, for Inupiat). Exceptions are myths, in which case the inhabitants cannot always name an associated individual. Here, East Greenlanders sometimes add a sentence such as 'this is what the people in former times said,' and legitimacy is created by pointing towards many retellings (cf. Mather and Morrow 1998, for a similar account from Yupiit in Alaska). Yvon Csonka writes that 'most Inuit make a sharp distinction in conversation between memories of facts that they have personally witnessed and those they have heard from others, and are often loath to report the latter' (2005: 325). Similarly, I have hardly ever heard people engaging in speculations about other people or occurrences without a 'proof' in terms of their own experiences, statements of the person talked about, or at least a statement of a third person who had personally witnessed that particular incident. Such a 'proof' in terms of stating from where or from whom one has attained certain information, is often demanded by listeners. Accordingly, I have frequently experienced Sermilingaarmeeq asking '*kia orarpa?*' (who has said this?). I will illustrate this with some examples.

One day at my host family's home in Sermiligaaq, Paula was telling her daughter Maline about a man from Tasiilaq who had recently died. Maline, at first, did not quite believe what her mother had been telling her, which she communicated through a sceptical gaze, a type of gaze which I have seen on many similar occasions. Only when Paula mentioned a credible source of information – in this case a person who had told her – did Maline believe her mother. Another similar incident took place at a good friend's house in Sermiligaaq where I was sleeping over one day. Shortly before going to bed, my friend took some painkillers. Since I had not noticed that she was not feeling well and was surprised by the dose of the pills, I asked her if she regularly took this quantity of paracetamol. The answer was '*kia orarpa?*' (who said that?). I replied that I did not know, but did not quite succeed in explaining that nobody else had been telling this to me, and that my question was supposed to be a rhetorical question. I realised that I was not supposed to ask such a thing without some kind of evidence. What I had not understood at that point was that such a question creates evidence instead of asking about it, as I had meant to do (see Goody 1978 on different functions of questions). And this creation of evidence, again, presupposes personal observations or at least to be able to refer to another person's statement, which I was not able to do.

These examples underline the fact that the Iivit, similar to the Yupiit in Alaska, 'see their society in terms of individualised positions, and not in terms of valorised norms' (Morrow 1990: 152). Shared expectations, which nevertheless exist, are communicated rather subtly and are not explicitly demanded. In their introductory speech at the Inuit Orality Conference in Paris 2008, Therrien and Collignon have underlined 'how embedded the Inuit sense of knowledge as individual practice remains, rooted in a rich unique lived experience' (2009). Likewise, Briggs speaks of an 'acceptance of personal truths' among the Inuit, 'all considered equally valid when supported by personal experience' (1998: 15). In view of this, Tunumeeq are very cautious both in speaking about another person's opinion or experience and in interfering with it by telling others what to do. These practices seem to imply an inappropriate intrusion on another person's autonomy. Similarly, I have rarely experienced Tunumeeq attributing any general statement to collective opinion or sentiment (except when referring to the past).[6] A few times when I asked a villager about 'the Sermilingaarmeeq' opinion I was told the person's own view and that she does not know about the others. This 'individuality' of knowledge in East Greenland has also been observed by the anthropologist Cunera Buijs (personal communication, 23 April 2007).

The above examples of reported speech and teasing illustrate the particular relevance of what another person has *said*. If a person says something, he or she has to take responsibility for the uttered words and for the possibility that others might pass them on. Statements made cannot be undone. I often had the impression that verbal utterances are perceived as particularly powerful in themselves, aside from the specific meanings of the words (cf. Holm 1911: 88; Flora 2012 on West Greenland, and Therrien 2008 on the Canadian Arctic). I therefore argue that we must look at words as producing effects rather than having or conveying intrinsic meaning.

In anthropology, this idea goes back to Bronislaw Malinowski's writings (1923, 1935) on the magical power of words, subsequently taken up by Tambiah (1968).

Malinowski argued that each utterance is essentially bound up with the context of the situation and the aim of the pursuit. He regards 'speech as a mode of action, rather than as a countersign of thought' (1923: 326). In his early writings, he highlights that 'words in their primary sense do, act, produce and achieve' (1935: 52), which he initially associates with their magical function in so-called 'primitive' language and societies (Malinowski 1923). Yet in his later works, Malinowski abandons this view and argues that there is not much difference between the 'savage use and the most abstract and theoretical one' (1935: 587). In linguistics, John Austin (1962) proposed a similar argument in his concepts of the speech act and the 'illocutionary' use of language. He distinguishes the act of saying ('locutionary act') from what one does in saying ('illocutionary act'), and what one does by saying ('perlocutionary act'). His notion of 'illocutionary acts' has been advanced by John Searle (1969). Like Malinowski, Austin regards speech itself as a form of action, though, contrary to Malinowski, he makes a clear-cut distinction between meaning and (illocutionary) force (Austin 1962: 100). In philosophy, Ludwig Wittgenstein (1953/2001) enquired into the relation between meaning and the practical use of language and famously compared words to tools in a toolbox, which may have varying uses. For him, language is like a game in which words may be used in a multiplicity of ways. The meaning of a word presupposes our ability to use it, and, Wittgenstein argues, 'meaning just is use' (i.e. words are not defined by reference to concepts but by the way they are used) (*ibid*: §77).[7]

This power of words (or to use Austin's term, their 'illocutionary force') has been mentioned with respect to many arctic peoples. Phyllis Morrow's (1990) account of the Yupiit of Alaska, for instance, is very similar to what I found in East Greenland. She explains the ambiguities and the general carefulness with regard to communicational and behavioural modalities in terms of this particular power or effect:

> At a deep level, I would argue, indirection and non-specify are outgrowths of a sense that speech and representation, along with vision and thought . . . , are inherently powerful. The power of words in marked contexts is very apparent.
> (1990: 152)

This power is not confined to speech but includes other sensory modes of perception and communication, such as vision, which I have touched upon in relation to communication with animals earlier in this book.

Moreover, the power of words, I believe, is also responsible for the fact that in my experience, deliberately lying – especially in face-to-face encounters – is very unusual in East Greenland. When asked a question one does not want to reply, the inhabitants often just say nothing, and in this way signal that they do not want to talk about a particular issue. This is accepted by the questioner, who usually does not pursue the matter. Kleivan has mentioned this point with reference to the song and drum duels from former times (see Chapter Six). According to her, these duels suggest 'that it was considered a breach of ethics to manipulate the facts and assert things for which there was no foundation in truth . . .' (1971: 11).

I have the impression that behind this susceptibility to the power of words lies something more fundamental about the East Greenlandic notion of the person. Many studies of personhood clearly divide an 'inner' self from the words it emits, the latter of which are associated with the (social) person, defined as the public image of the self. According to this conception, selves do not impinge on one another directly in speech, but only indirectly by way of the words released into the interpersonal domain of interaction. For the Iivit, however, there seems to be no such separation between an 'inner' self and a social person situated within the domain of interpersonal relationships, a point that I touched upon briefly when speaking about the continuity of personhood in relation to naming practices in Chapter Two. Selves, rather than being sheltered within bodies, appear to leak out into the world on the current of words (and, of course, of other gestures and bodily practices), which renders them extremely vulnerable. Selves are not separable from social persons, or from particular roles (*personages*), or masks (*personas*) (a distinction coined by Mauss 1985); they have no outer, bodily shell to protect them, and, through their words, they can directly impact one another. There is no hiding place – no private domain for the person to inhabit – that is set apart from the public domain of interaction; the world is interpersonal through and through. As manifestations of being, words give you away just as facial expressions do. This might be the reason why people have to be so careful in what they say, and why they cannot lie about things.

This inseparability of the self as a psychological unit and the person, in the sense of a social or public being, has also been found for some other hunting and gathering societies, such as the Ilongot of the Philippines (Rosaldo 1980, 1984). Among the Ilongot, quite similar to East Greenland, the spoken word has a strong impact, and, as Rosaldo explains, it is therefore dangerous to show anger and to reward offences – once expressed, such anger is manifestly present and cannot be undone. Also among the Inuit, as has been shown most notably by Briggs in her famous book *Never in Anger* (1970), anger is not usually directly expressed, but is disguised by various subtle and indirect ways of communication. I shall now return to some further examples of interpersonal communication among house-mates and family members, which illustrate the power of words and the high value placed on personal autonomy.

Expectations of giving information, questioning, and not-answering

Daily life in and around the house is characterised by regular comings and goings. When family members entered or left the house, I found that it was often not appropriate to ask a person what she had been doing or what she was planning to do. Short questions on where somebody is coming from (*sumingaanii?*) or where she is heading (*sumut?*) are posed rather regularly, but they would often be answered rather concisely. Sometimes East Greenlanders just reply with a vague arm movement, indicating a direction, or naming a destination or place for which they were, or are, heading. At other times, people say '*silamut*' or '*silamingaanii*' (I am going/

coming from outside), which can mean anything from meeting people outside in the village to a stop-over at some person's home.[8] What information is provided seems to be very much up to a person who decides him- or herself if, when, and about what he or she wants to inform others. I rarely encountered further requests for specification from other people, and this seems to be an improper intrusion into another's personal sphere. Yet, asking about the reasons for leaving somebody's company is quite common when visiting other people's houses, as I have mentioned above. In this case, I often found that people from the visited household expected a reasonable explanation of the particular activities or obligations that drew the person away, at least if these reasons were not well known or obvious.

For example, when a family member returned after a day spent at work in the village or from a trip, he or she was usually not asked about the happenings of the day by other household members, though I knew that sometimes housemates were keen on being informed. In the following example I will illustrate a common situation, which I have often encountered in this or in a similar manner. A man enters the house after a day out hunting and fishing, leaving the catch and sacks of fish on the porch. He is tired and needs some rest. After having fetched himself some food from the kitchen and having eaten at the dining table or on the kitchen floor, he establishes himself on the sofa and watches television or plays with the children. He does not speak about the happenings of the day. Only after some period of time, which might be up to some hours, he rather casually drops a comment about how many seals he has caught or the like, and begins to elaborate upon the happenings of his trip – issues that are all of particular interest to the cohabitants. Only in few cases I noticed that a wife or other family member would question the husband at an earlier point. Usually they would wait for his initiative. In this and similar situations, other household members are expected not to show curiosity and to wait for the moment when the person feels like talking.

Similarly, children are taught to behave in this manner, to refrain from asking and to hold back curiosity. For example, one day we were all sitting together around the dining table taking supper at my host family's place. We had just started to eat when Lars, the head of the family, got up. He mumbled a short comment and went out of the door. I picked up the word 'police' from the children talking next to me, which led me to assume that Lars's sudden departure might have to do with his work as a police helper. Then the children asked their mother Maline about where their father went. She succinctly answered '*nalivarnga*' (I don't know) and that he has been 'called'. Through the way she said this, she signalled to the children that they were not supposed to go on asking. With a questioning look towards Maline, I dropped a comment suggesting that his departure might have to do with some drunken people causing trouble. Maline raised her eyebrows (thus signalling yes) but would not further talk about it. Paula, the grandmother, mentioned the name of a person who could be involved. Later on, when Lars got back, nobody asked him where he had been. The children (and I) had understood that they were not supposed to show interest in this issue. Thus, in this case, the principle of not approaching Lars, on the one hand, rested upon his specific task as a police helper, which involves activities that are not supposed to be talked about, especially in front of children. On the other hand, it also illustrates a person's capacity to determine for him- or herself whether to volunteer information, according to context. All in all, I found that East Greenlanders are often rather careful in requesting information from one another and often wait for a sign that the person in question wants to tell something before starting to ask. As Morrow has written, these examples illustrate that 'choice of silence or speech belongs to the individual' (1996: 414).

However, I was sometimes rather surprised by Tunumeeq's upfront questions on issues that I myself would not have dared to ask about (at least not in the initial stages of fieldwork). These questions, for example, were about illnesses or particular bodily features or functions. In this context, Sermilingaarmeeq did not hold back in articulating their opinions. For example, I was told that my freckles are not very pretty, but my face looks nice. Likewise, some people remarked that they did not like the short hair covering my forearms. More generally, villagers often comment about other inhabitants' outer appearances, such as when a person has lost or gained weight, smelled, wore dirty clothes, and so forth. These incidents differ from the examples of 'individual knowledge' I have given above, in that they concern things that are visible or perceivable to others, providing a shared context for participants in conversation. These conversations are thus not based on the knowledge of one of the participants only, such as in the case of asking a returning hunter about the happenings of the day. The shared context is used as a starting point for conversational exchange, in much the same way as with conversations sparked off from watching television.

Moreover, these examples of posing and answering questions illustrate not only patterns of talking, but also the use of silence, that is, of not answering. Silence, in this case, does not mean that communication is broken or interrupted. On the contrary – these quiet moments are themselves a meaningful way of communicating that a person does not want to speak about a particular issue. Hence, it is important to keep in mind that silence has similar communicative functions to speech. As Jaworski writes, 'the main common link between speech and silence is that the same interpretative processes apply to someone's remaining meaningfully silent in discourse as to their speaking' (1993: 3). Following Wikan, I argue that we have to regard 'silence not as a void or absence or emptiness but as a fill and pregnant with meaning' (1993: 201).[9]

The importance of silence became very apparent in East Greenlanders' ways of teaching and enskillment, such as in the context of the education of children. Apart from observations on caregiver-child interactions, some of the following examples also stem from my personal experience of being an apprentice in various tasks of daily life in Sermiligaaq, a role which partially corresponded to that of a child undergoing socialisation.

Learning processes and the education of children

Children in Sermiligaaq, and in many other Inuit communities, receive a lot of attention and affection in daily life. Members of the children's families, just as many other people encountered in daily life, frequently give them attention, and play with them, cuddle them, and interact with them. Many anthropologists have remarked that learning among arctic people is achieved through observation and practice (e.g., Bodenhorn 1997; Briggs 1998). This also holds with respect to my observations in East Greenland.

A child is called *mertserteq* in East Greenlandic, in contrast to the term *naaalivi-arteq* which is used for a baby until it is able to walk. Being present at many adult interactions as silent but attentive observers, children learn through looking and paying attention. For instance, children frequently accompany their caregivers to all sorts of community events such as political assemblies, meetings of associations/societies, and so forth. During these events children may move around relatively freely, and I have rarely heard a person complain that they would disrupt the adults' activities. These activities also include so-called traditional activities such as hunting and fishing as well as tasks taking place at people's homes, whereas they do not include wage labour activities as these usually do not allow for the presence of children. Initial phases of observation are followed by supervised participation and a period of unsupervised self-testing (cf. Philips 2007 on learning among Native American children of the Warm Springs tribe). The use of speech is minimal, and communication often takes place through showing and looking. Working together is often characterised by a quiet atmosphere and a pleasant rhythm of conjoint or alternate activity of the skilled practitioners (i.e. parent) and the apprentice (i.e. child).

For example, during the short winter days in 2007, Sibora taught me how to knit socks. In doing so, she never explained the exact procedure to me as I had initially expected (and a few times I was almost disappointed by that). What she did was either take over the knitting, correcting my mistakes and continuing the work whilst I was watching, or give the knitting back to me, not paying further attention and letting me try by myself, whilst continuing her own work. Talking in these situations, apart from some shorter comments, was not much appreciated. So, either I found myself in the position of a silent observer, or I was put into a self-testing situation. Never did we go over the knitting work together, nor would Sibora verbally explain every step.

Children's undertakings to try out particular skills and practices may at times involve making mistakes, or even lead them to have rather unpleasant experiences. On such occasions, family members would intervene only if there was danger of serious damage or hurt. For example, the little boy Nuka at my host family, who was one and a half years old when I first came to Sermiligaaq, would not stop reaching out for the hot plate in the kitchen for some weeks, though he was told '*agaa*' (this hurts) a number of times. One day, he slightly burned his fingers, which was not prevented by family members nearby. After this somewhat painful experience, I did not see Nuka trying this again. A similar instance has been documented in the documentary film that Anni and I made (Seitz and Elixhauser 2008). Sitting on a small stool on the kitchen floor, Maline was cleaning a sealskin. Thala and Gerth, who were three and seven years old at that time, were climbing around the kitchen furniture above their mother's head. The workspace was covered with kitchen utensils and dinnerware. At one point, Gerth stumbled across a plastic cup; a little bit later Thala almost touched the blade of a kitchen knife she was walking across with her bare feet. Maline merely remarked: *'Qaqqasiartattu?'* (Are you climbing?). For the rest, she did not seem to pay attention and concentrated on her work. Just a few minutes later she told Gerth in a soft and warm voice, *'ernilarnga uninngaqqarniatsilit'* (my son, you better stop with that). Shortly after, the children left the kitchen.

These examples illustrate the general observation that, similar to what a number of anthropologists have remarked with respect to other arctic peoples (e.g., Bodenhorn 1988, 1997; Briggs 1991, 1998), it is not very common in East Greenland to scold a child or to give explicit verbal instructions on how to behave. Thalbitzer has stated, referring to observations from East Greenland during the first half of the 20th century, 'The Eskimo does not punish his child' (1941: 600). This, along with a very similar remark by Gustav Holm (1911: 63), is – with some exceptions – still more or less applicable to the contemporary situation, at least for what I have encountered in Sermiligaaq and for some families in Tasiilaq and other settlements in the Ammassalik region. Children are not prevented from making mistakes and, moreover, are often left to resolve quarrels without the interference of their parents. My observations support Briggs's assertion that 'Inuit education . . . is a process in which children actively experience and independently try to solve problems that are the substance of their daily lives, both as children and as adults' (1991: 282). Hence, a child, just as any other person, is perceived as an autonomous agent who is capable of taking decisions by him- or herself (*nammeq*), and is often asked about his or her own opinion (Gessain 1969: 35–6; cf. Bodenhorn 1997; Briggs 2001; Flora 2012).

This becomes apparent in the opening anecdote in the introduction of this book, in which I had asked my friend Maline if her daughter Thala would be going to the kindergarten that day. Maline had passed the question on to the three-year-old girl, who was to decide herself (*nammeq*). Grete Hovelrud-Broda gives a similar example from her fieldwork in the village of Isertoq in the south of the Ammassalik district. A resident once told her: 'You must not ask me what that little child wants to do. How can I know what someone else wants?' which, Hovelrud-Broda explains, shows the 'strong belief in the individual autonomy that is deeply embedded in the East Greenlandic culture' (2000a: 154). Likewise, my friend Hansigne explained to me that her daughter had stopped going to the kindergarten after a few times, since she did not like the other children. This pronounced right to choose for oneself also showed with respect to the food the children eat as well as their eating times. In the East Greenlandic families I know, children usually eat when it suits them best and are relatively free in deciding on the kinds of food they prefer. In my host family, for instance, as in other households I know, the children are allowed to help themselves from the fridge and the kitchen cupboards. Only a few items are excluded, since, as Maline once explained to the children, they 'belong' to one of the adults. The children eat when they are hungry, when they get back home after school or after having finished playing outside. Eating times, thus, differ from day to day. Moreover, I never experienced any pressure on children having to finish a plate of food. Besides these examples of issues that children decide for themselves are a variety of others, up to questions on where a child wants to sleep, or in which

Figure 4.5 Sermiligaaq girls in front of the meetinghouse, January 2007
(Photo: Sophie Elixhauser)

household it wants to live. Bodenhorn reports, with respect to Alaska, that even in cases of adoption a child may sometimes decide for him- or herself (1988, 2000a). I do not know of any similar case from East Greenland, but I can very well imagine that children's voices may be decisive in the matter.

This autonomy might also be related to naming practices in East Greenland, which as explained above imply that when a person dies her name-soul is passed on to a newborn baby. Other people recognise characteristics of the deceased person in that child, which in some way becomes the deceased person, as East Greenlanders have told me. The child also takes over the deceased person's kinship relations, and is often addressed with the same kinship terms. In the light of these practices it becomes understandable that adults show great respect to children and hardly ever scold a child, who for them might, for instance, be their grandmother or some other relative from whom the child carries the name.

Nevertheless, children in East Greenland are not generally free in deciding about all kinds of issues that concern them, and they have to follow their parents' or other family members' expectations once in a while. For example, a parent might decide if a child may accompany him or her to visit somewhere, or if it has to stay behind during boat trips or longer absences; parents tell their children to do their homework, to be in time for school, or to brush their teeth. These explicit rules and guidelines are often (yet not in all cases) related to one or another of the official regulations that structure modern life in a Greenlandic settlement. Nevertheless, such regulations are not always strictly enforced by the parents, and sometimes orders are formulated as suggestions instead of telling children what to do (cf. Robert-Lamblin 1986: 145).

With respect to decisions about a child, the strong and influential role of the grandmothers caught my attention. They often assert great influence in the upbringing of their grandchildren and are involved and sometimes decisive in issues such as, for example, when deciding if a young woman should breast-feed her child or not,[10] whether a pregnant family member should get an abortion, or where a child should be living. Likewise, also among couples, wife and husband (or unmarried partners) sometimes tell each other what to do, in various subtle or not so subtle ways. Thalbitzer mentioned that, in addition to 'circumscribed individual freedom', a certain authority was held by the elders of the house and the male family head (1941: 625). These observations are somewhat at odds with the principle of not interfering in other people's decisions and illustrate that the two values of autonomy and responsibilities in social relationships sometimes conflict.

Morality play

Apart from the few explicit rules and regulations, children are confronted with a variety of expectations that are communicated in a rather implicit way. Some of these are introduced through stories, others through a certain way of joking and sarcasm, or *morality play* as Briggs (1998) has called it. In the following section I will elaborate upon the latter. Ann Eisenberg has argued that teasing creates a safe mode to 'communicate messages that might have been inappropriate to communicate directly . . . This type of pointed teasing works for the teaser because he or she can always deny any intent to convey a message' (1986: 190). It therefore allows people to indicate right conduct or to bring disapproved behaviour to attention, without explicitly having to state it. Learning to understand these subtle messages, however, is not always easy for a child. Eisenberg continues,

> teasing is . . . a linguistic skill that children may have to learn to manipulate to speak like the adult members of their particular group. Learning to participate in teasing and to recognise that one is being teased requires sensitivity to nonverbal cues and an ability to go beyond the surface meaning of a message to determine the intentions of the speaker.
>
> (*ibid*: 196)

Teasing is not just a speech genre, but also involves a variety of other communicational modes such as to be able to understand intonations of voice, facial expressions, accompanying gestures, or bodily postures. To learn to understand teasing and to properly reply, moreover, requires the knowledge and experience to be able to assess if a statement could be true or not. The Inuit type of play builds precisely upon these ambiguities and on the fact that children do not always know what is serious and what is not.

The particular way of teasing that I regularly encountered in interactions between East Greenlandic caregivers and children involves an adult (or older child or teenager) saying the exact opposite of what she means, and thereby testing the child's knowledge and ability to recognise this as playful. If a child agrees, the adults react with laughter, whereas a child's disapproval is met with positive reinforcement. This verbal play most often involves toddlers and smaller children, as the older ones have already learned to be careful in agreeing to such offers. In such a 'test' situation, adults may use a number of verbal tricks to try to make the child agree to such an 'opposite' statement. For instance, I experienced a number of times when the child was asked to repeat an affirmative sentence which, in case it did, was then picked up by the adult and held against the child. An example will illustrate the basic principle of such play. One morning in Sermiligaaq, Samuel, the grandfather in my host family, had left very early for a hunting trip of several days. His grandson Nuka (two years old) asked '*Naak aalaq?*' (Where is grandfather?). The mother Maline teasingly told him that grandfather left for the town of Tasiilaq, located in the opposite direction from where Samuel went. 'Yes, in Tasiilaq', she assured a few times. Nuka felt that this could not be right, which you could see from the sceptical look on his face. He did not say anything though, and appeared somewhat lost in trying to establish whether his mother was telling the truth or not. The other people around started to laugh. Nuka remained rather indifferent. Maline, thus, tested the boy's ability to realise that she was joking, and to pay attention to particular indicators that had pointed towards the grandfather's actual destiny such as the preparations he had taken the night before.

Another common theme of such play, which has also been reported by Briggs (1998: ch. 4), calls into question a child's household membership. I experienced a number of times that children were asked, half-seriously, if they wanted to accompany a departing guest and to continue living at that person's place. The guest would try to talk the child into it through mentioning all sorts of advantages his or her household had to offer. The child then had to decide how to react and if it wanted to approve or not. If it had not understood the subtle signs of irony and other clues, it would approve, which would be met with teasing and laughter by the adults. They might ironically ask the child why it had agreed to leave its parents and family through such questions as: 'You want to leave us?' 'Don't you love us?' 'Why would you leave us?' Thus, through dramatising the consequences of an answer, adults teach the child important lessons, which in the case of this example concern belonging and the value of home and family. I have observed similar plays with respect to various topics, from harmless versions such as the example about Nuka's grandfather to scenarios of being abandoned or rejected and leaving a child behind. The latter version calls the children's attention to personal weakness, the dangers of being on one's own, and teaches them the importance of social relationships. Furthermore, such play teaches children that they are expected to draw their own conclusions, and to think for themselves. It supports the development of independent problem-solving skills, and makes children attentive to various subtle and not so subtle clues that help to disentangle the playful from the serious. Such clues include signs of irony, tone of voice, facial expressions, and the like, as well as recurring key phrases and questions (cf. Briggs 2000b, 2001).

On a broader level, East Greenlandic children are frequently the butt of jokes and laughed at by adults, be it for clumsiness in daily life or a funny statement they have coined. Situations of being teased can at times be pretty tough and demand a great deal of self-control, as I have also experienced when being teased myself. In this respect, it struck me that I hardly ever encountered children (or other persons) showing emotions when being teased or laughed at, and seldom would a child start to cry. I believe this can be explained by the fact that, from an early age, children begin to get used to these situations, with their inherent tensions and ambiguities. They learn that they can make adults stop teasing and thereby protect themselves through not showing that they are bothered. Inge Kleivan also mentioned this emotional control in an article on drum and song duels in former West Greenland. She explains that opponents in such a contest with 'underlying resentment of varying intensity, were expected to act smiling and unconcerned' (1971: 17). Mahadev Apte formulated the alignment of teasing and emotional self-control. He writes that adults 'tend to tease children deliberately in response to distress or anger, believing that teasing will teach the youngsters self-control, especially if society emphasises emotional self-control in social encounters' (1985: 101). In this I have made very similar observations to Joëlle Robert-Lamblin. She remarks that the 'control of emotion, reaction or gestures that might lead to a loss of the *isima* (mind, thought, that which makes one human) is taught from infancy through the family upbringing and strictly enforced by the family group' (1986: 145). This control of emotions has been noted with respect to many Inuit peoples around the circumpolar North, as in Briggs's ethnography *Never in Anger* (1970), or Deanna Paniattaq Kingston's (2008) reflections on the persistence of conflict avoidance among the King Island Inupiat. Children's reactions to being teased are thus closely related to the avoidance of confrontation and the expectation of not imposing one's own opinions on others, such as through giving orders. In this context as well, Iivit usually do not show that they are bothered and counter with indifferent facial expressions, withdrawal, and sometimes irony. Another statement by Briggs very well describes these reactions: '"Stable ambiguity" was the ideal: no winners, no losers, no innocent and no guilty parties.' (2001: 236)

More generally, with respect to teasing among family members, kin, and friends, one encounters particular joking relationships, i.e. joking and teasing between certain pairs or groups of people who are not only permitted, but also expected to behave in such a way (see Radcliffe-Brown 1940, 1949; Apte 1985: ch. 1; Kleivan 1971 on West Greenland). I found such relationships in East Greenland, for example between potential marriage partners and people from alternating generations. It is important to note that such relationships 'serve both to avert open conflicts and to strengthen friendships' (Kleivan 1971: 17). They stabilise relations that might be subject to conflict and at times intensify friendships (cf. Eisenberg 1986). Eisenberg explains: 'That it is "safe" to tease a particular individual indicates that a special relationship exists' (*ibid*: 193). Teasing, thus, presupposes some kind of familiarity and relationship between those involved. Not to be teased any longer is indeed a clear sign of dissonance, conflict, or that something is wrong, as I have experienced a number of times with respect to misunderstanding, in which I was personally involved.

I now want to return to teasing in interactions between children and adults. In this context, too, play frequently involves one's own body, not only for communicating an issue through gestures and the like, but also when playing with it, joking about it, and using it as 'illustrative material', as I have shown above. The relationship between caregivers and children is characterised by close physical contact, and children are shown great affection through cuddling, hugging, and so forth. This physicality, I believe, influences East Greenlanders' ability to read and understand many little behavioural details that are communicated through speech as well as various bodily movements and expressions. Following Briggs, I argue that

> this constant physical-social contact with others as well as constant use of their own bodies as objects, and practice in active imitation help to make even very small children, not to mention the caregivers, acutely observant of the ever-changing details of their own and other people's behavior.
>
> (1991: 274)

Physical contact and affection become most apparent in people's homes, an important locality for intimate relations among the people inhabiting this space. When spending time together with the family in the evenings on the couch, for instance, Tunumeeq often sit closely together, not making use of all the available space. Such moments within the circle of close family members provide occasion to communicate rather intimate topics, through various verbal or nonverbal means. Yet, in situations of spending time with family members within the limited space of the house, it struck me that sometimes individuals very obviously withdrew from communicating with one or several people, even though they stayed in their presence.

Involvement with cohabitants and personal space

As I have shown, the people in Sermiligaaq and other East Greenlandic settlements often live together in comparatively crowded contexts and do not retain much personal space. In addition, many Iivit I met prefer being in the company of other people as much as possible and only exceptionally seek solitude. Nevertheless, at times individuals decide to withdraw from those around them and not to interact with (particular) people. This, most generally, may take place through leaving a room, going to visit elsewhere, or, in some cases, through moving away, as I have described earlier in this book. However, such a retreat does not necessarily entail a spatial separation from other people. I have encountered such suspension of interaction with others while staying in the company of people in a number of different ways (categories which certainly do overlap): first, persons may decide that they do not want to interact with others or with a particular person for personal or other reasons; second, they may be the butt of such practices and be ignored by another person or a group of people without actively striving for this position; or, third, they and others may not interact with each other due to socially accepted conventions relating to gender, professional or other types of discretion, and so forth. I will illustrate these with some examples.

When gathering with other household members and visitors in the living room during evenings, it sometimes happened that individuals signalled that they did not want to be included in conversations or to receive attention. This was communicated through showing no interest in other people, by averting the gaze, and by not reacting to what was spoken and happening around them. These individuals suspended most interaction with the surrounding people, but at the same time stayed in their company. Possible reasons for such retreat include that the person was tired, was concentrating on something else such as the television or a skilled activity, or was busy with his or her own thoughts or emotions. This habit of 'cutting off' involvement[11] was accepted by surrounding people, who did not usually try to approach the person in question or enquire about possible reasons. This would have meant an inappropriate intrusion. Likewise, I was often surprised by East Greenlanders' ability to successfully 'not notice' another person, acting just as if that person was not present, whilst behaving quite normally towards everyone else (see Hall 2003 on proxemics, and cf. Haviland and Haviland 1983 for a similar account from Mexico). I observed this with respect to personal dissonance or tensions between individuals, in which cases the persons in question would try to avoid each other's company, and sometimes simply disappear – from a room, house, area, or settlement. However, sometimes this spatial withdrawal was not possible or wanted.

For example, I once joined a family boat trip to another village for some days, together with a friend, a foreigner like me. During these days, the wife of the boat driver did not pay much attention to us. Quite obviously there was jealousy involved, as the wife did not like the way her husband paid attention to my friend. She thus avoided social interaction with us just as did most of her relatives who joined the trip. This became most apparent when we were all on our way, crammed in the boat, and also later in the evenings in the house where we were staying. Rather, as Briggs has written in her book *Never in Anger* (1970) about her being ostracised and how she did not at first even notice, this trip led me to experience how it feels to be close together with a number of people but not to be integrated in conversations, joking, and the like, almost as if one were invisible.

During my time in Sermiligaaq, I was especially struck by a different way of understanding privacy in relation to particular rooms. For example, locations in the house frequented by many people and freely entered by everyone, such as the living room, were sometimes used for personal conversations between two or more specific inhabitants or visitors. These conversations took place whilst other household members stayed in the same room and were potentially able to hear and see what was going on. The people around were not supposed to pay attention to what was being talked about, but they were also not expected to leave the room. I have experienced such instances a number of times, sometimes as an observer, and sometimes also as a participant. In most such cases, expectations of not paying attention related to factors such as gender, personal intimacy, and professional discretion.

I will give an example. One afternoon I was visiting at a friend's place in Sermiligaaq where I was a regular guest. I was sitting on the sofa studying my vocabulary book. A man from the village entered the house in order to see my friend's husband. The two men sat down on the other side of the sofa where they had a lively conversation. It was clear that the men were talking to each other and that the conversation was not meant to include the children in the room or me. I went on studying my vocabulary. Apart from a short moment when the visitor asked me a few questions, I did not participate in the conversation. Yet, because of the spatial proximity, I overheard snippets of it. My female friend, the house-wife, was in the kitchen for most of the time. On a few occasions she came into the living room and stood next to the men, listening for a little while and giving some comments. After the visitor left, I was doing the dishes with my friend in the kitchen and asked her about some issues I had picked up from the men's conversa-tion. I thought I had heard something about a theft that had taken place. But my friend explained that she did not know, as this had been 'the men's' conversation. I had observed that she did participate in some parts of it, but she told me that the conversation should not interest us.

This example shows that if something is being discussed in the living room, or also, for instance, in a communal building, this does not automatically mean that everyone present is supposed to or allowed to listen. In this case it was a question of gender. As I have mentioned above, there are rather strict expecta-tions of how to behave towards non-related persons of the opposite sex. I do not understand all the details of this, and as an unmarried woman doing fieldwork my experience has been rather slanted. Nevertheless, I often found that I was not supposed to talk to or to pay attention in one way or another (through speech, gaze, or other kinds of communication) to men I met, especially if they were not members of families I knew well. I believe that similar reasons are responsible for the fact that the man who came to visit my friend's husband hardly took any notice of me.

Another example will underline this argument on gender, spatiality, and paying attention to other people. One day I was visiting the house of a friend in Tasiilaq. Her family and I were hanging out in the living room, watching TV and chatting. I was sitting next to her teenage son on the sofa, and the two small daughters were playing all around the room. I did not pay much attention to the son and vice versa. Verbal exchange was mostly channelled via my female friend sitting on the oppo-site side of the corner sofa. After a while, the two girls sat down on my right hand side, while the son was still sitting on my other side. I was playing with the girls whilst the boy was concentrating on the television programme. The girls next to me were not part of his field of vision any longer. At one point, however, the girls were getting a bit too wild pulling at my hair, and the son right away told them to stop. Though he was still looking at the television straight ahead, he had picked up every detail of his younger sisters playing on my right-hand side and me. Also whilst telling the girls to stop, he was still looking in the direction of the televi-sion. I was surprised that he had actually been able to notice the details without turning his head. That evening I noted in my diary that it came to my mind, once again, how intensely Tunumeeq are able to *not* look at other people while still taking in a great deal of detail from the surroundings, just from the corner of their eyes. Obviously, it would not have been appropriate for my friend's son to turn his head towards me (and in the direction of his sisters), as this would have signalled some kind of interest in me as a woman. Yet, just like hunters, and likewise many sports practitioners (cf. Downey 2007), he was able to pick up movements across a wide field without being distracted by objects in his focal area. The young man was looking at the television, and yet he was aware of the movements of his sisters through his highly trained peripheral vision. This example once again illustrates the particular power of focal gaze, the meanings associated with it, and the importance of using peripheral areas of the visual field. The latter appears especially trained among the Iivit.

The illustrations I have presented have shown that paying attention or not paying attention to another person is something to be learned. Accordingly, 'ignoring' means actively disregarding somebody, which in the East Greenlandic context can relate to ways of dealing with emotions, personal dissonances, and conflicts, different kinds of prohibitions or confidentialities, gender relations, and so forth. These observations highlight our ability to separate the persons we consider fully-fledged participants in given situations from others whom, although they are physically present, we ignore (Zerubavel 2006: 24).[12] These examples of personal space and of varying involvement with other people indicate the fact that what people see, are supposed to see, want to see, and so forth are all matters of how to direct one's attention. Cristina Grasseni explains vision as an 'embodied, skilled, trained, sense that characterises (certain) practices' (2004: 41). She writes that in order to learn to visually perceive in a certain way, one needs education or even training in a relation of apprenticeship. From my personal experience of fieldwork in East Greenland, I can confirm that this included a personal enskillment in how to look and visually perceive other people and things in my environment. This attention, moreover, implies a certain 'tuning', an intuitive grasp of the body of the other, which is furthermore crucial for every infant's original corporeal-kinetic apprenticeship. Maxine Sheets-Johnstone (2000) notes that this 'tuning' also happens with regard to other persons: eye-to-eye-contact, for instance, is an experience of attunement, just as is communication through sideways glances. This particular education of attention happens not only in relation to the perception of surrounding people, but also when children learn on which occasions and about which matters they may and may not ask or show a certain interest, and also with respect to the issue of curiosity.

These examples have illustrated that children learn to pay attention to a wide range of communicative modalities and indicators. These include not only subtle clues, irony, and the like, such as I have mentioned with respect to the playful interactions with adults, but also culturally appropriate rules to actively ignore particular issues, i.e. not to listen to conversations or not to look at other people. Very similar rules become apparent when looking at how the house, with its windows and doors, structures communication, for example with respect to residents' use of particular rooms that I have shortly touched upon above, and also with respect to communication between people inside and outside of a house, to which I will turn now.[13]

The crannies of a house: curiosity, seeing others, and being seen

East Greenlandic houses are usually pretty light due to their several windows (*ingalat*). The windows are often decorated with plants. House plants, which are imported and sold at the local stores, are very popular in East Greenland. Most women pay considerable attention to curtains, tablecloths, and adornments, such as little figures, clocks, plants (including plastic plants), and framed pictures, which also cover windowsills and panes. Christmastime is the peak of the year in this regard, and most homes are packed with all sorts of colourful Christmas decorations. Many households, for instance, put up a big orange Christmas star in one of the windows that is lit with a light bulb and can be seen from afar (see Figure 4.4).

As with other settlements in the Ammassalik district, the terrain of Sermiligaaq is very hilly. Thus, most houses in the village are situated in locations that allow residents to observe the village centre, the fjord, or the harbour from one or several of the windows. Indispensable to sills of such windows is a pair of binoculars (*qernilit*), and Sermilingaarmeeq often make use of them in their regular lookouts (see Figure 4.6). They watch out for family members returning from school or the work place, or for the husband or other male family members arriving from hunting or fishing. In addition to searching for particular persons, inhabitants entertain

Figure 4.6 Dina observing her uncle returning with the dog team, April 2007

(Photo: Sophie Elixhauser)

themselves by 'binocularising' the important places in the village, and the activities and whereabouts of other inhabitants. These practices provide information about other Sermilingaarmeeq, people arriving to and leaving the village, as well as various other goings on.

Nevertheless, such practices of curiosity are not much talked about with people outside their homes and often happen rather secretly. Residents usually try their best to ensure that, apart from their family members, other people do not notice that they are watching, either through peering from an angle or supported by curtains, plants, and other items that help residents to simply disappear behind the panes. Nevertheless, if, once in a while, a person can be seen in the window from the outside, people passing by would often act as if they did not notice. For instance, a number of times I was walking by some villagers' houses in the company of Greenlandic friends, and we were able to see somebody standing at a window watching what was happening outside. My Greenlandic companions would not pay attention to this person, neither looking at him or her nor talking about it. A person standing in the window might now and again be cradling a baby in his or her arms. I have the impression that this gives a person a better reason to be candidly looking outside, as the responsibility for wanting to observe what is going on outside is turned over to the baby.

In contrast to their internal decoration and the life that goes on inside them, from the outside East Greenlandic houses often appear rather lifeless and void. On some days when passing through Sermiligaaq and not being familiar with the place, one could get the impression that nobody is at home. Yet among the Iivit, it seems to be common knowledge that houses have crannies that are used for looking out on what is happening outside, though this is not usually made explicit verbally. Sermilingaarmeeq frequently discuss issues that they have found out in this way, but I have not often heard people talking about the particular circumstances of how they had come to observe the incidents related; nor were my friends particularly keen to talk to me about them. For instance, when I once jokingly remarked that there are so many interesting things to be seen through the windows using the binoculars and that this reminds me of watching TV, my companions did not pick up that joke or laugh with me about it. Another incident will exemplify the same point. During winter 2006–07, I was trying to learn the Greenlanders' way of successfully walking on all sorts of snow and ice-covered grounds. I did not always succeed and slipped a few times in the snow or ice. One day, whilst walking from my host family's house to the service house, I again lost my balance and landed in the snow. I did not see anybody around, and at first felt lucky that I had not been watched, or so I thought, since I suspected that this would have resulted in more teasing on the part of the villagers. Nevertheless, this first impression was wrong. Indeed, a number of villagers had observed the incident, and over the following days I was asked and teased about it several times. These villagers mentioned a number of different names of people they had heard the incident from. Since I did not see anybody outside when it happened, I came to assume that these persons must have observed me through the windows (though no one admitted this).

Building upon several similar experiences, I found that instead of having to back up statements or stories by mentioning another person who has told you, the circumstances of how a person has directly observed a particular happening do not have to be evidenced. Exceptions include situations when the persons watched are part of the same family or household, and when the activities observed directly concern the 'curious' person. Then, the latter would tell the former that she had seen her or him through the window, such as in case of a wife who has 'binocularised' her husband way ahead out on the fjord before his arrival in Sermiligaaq.

In these observations, I find some parallels to what Georg Henriksen (1993) has written about the Naskapi in Canada. He describes the Naskapi as very curious with regard to other members of the group. Similar to what I have written about situational leaders and followers in Chapter Three, leadership among the Naskapi arises in proportion with ability. Asking a person if he or she plans to go hunting is often answered only by a 'perhaps', which is also quite similar in East Greenland. Among the Naskapi, too, one encounters a strong sense of personal autonomy that in some situations does not allow for too much questioning. With respect to leading other hunters, Henriksen reports that sometimes it is generally known who will be the leader, but at other times the situation may still be unresolved when the morning of the hunt arrives. The Naskapi are therefore very attentive to all kinds of little signs that might reveal who plans or hopes to be a *wotshimao* (situational leader) and whom they should follow (Henriksen 1993: 40–54). 'Therefore, in any Naskapi camp, people are continually "spying" on each other, trying to find out what everyone is thinking of doing. Even when sitting inside their tents, they spy through a hole in the tent' (1993: 48).

Thus, as in East Greenland, where in many situations a person's autonomy does not allow for frank curiosity, observing or monitoring others provides an important way of obtaining information on other people. These examples of watching other villagers through the crannies of a house show that there are particular expectations concerning what you may observe and to whom you may admit your curiosity. This is closely related to what I have written above, about situations of being in the same room with a number of people and not being allowed or not wanting to pay attention to particular people.

Houses, and some of their functions and uses, have sometimes been compared to faces, or to persons as such, and have likewise been considered as extensions of a person (Carsten and Hugh-Jones 1995). These comparisons are apt for houses and interpersonal communication in East Greenland, especially considering my treatment of curiosity. Thus, windows could be compared to eyes, and the walls of a house to the skin of a person. This is my own interpretation, as I have not heard East Greenlanders speaking about houses in this way. Nevertheless, Therrien (1987) argues in her ethnolinguistic study of the Inuit body that the habitat of the Inuit opposes interiority and exteriority. She supports her argument with linguistic and symbolic correlations between the traditional Inuit snow house and the human body, and their respective openings, as for example the associations of a spiracle with a window. Bordin (2003) argues that these correlations also apply to contemporary houses that have been assimilated linguistically to the traditional model.

He refers to Canadian Inuit, yet the East Greenlandic language shows similarities in this respect.

Similar to the many sensory-kinetic ways of perceiving your environment, including other people, a house offers greater 'permeability' than appears at first sight. As Carsten and Hugh-Jones write, 'like an extra skin, carapace or second layer of clothes, it serves as much to reveal and display as it does to hide and protect' (1995: 2). Moreover, practices of looking out of the windows parallel practices of seeing from 'the corners of the eyes' that I have mentioned above. In this respect, one encounters a difference between 'binocularising' other inhabitants and happenings from a distance and peering outside to observe people in closer vicinities. The former, most obviously, involves focal, focussed vision, but it is not reciprocated; the people watched are not aware of the observer. It is different, however, when a person observes another person passing by his or her house. Here, both parties may be aware of each other's presence, since on occasion the person passing by also notices that somebody is looking out of the window. Nevertheless, the former is not allowed to avert her gaze towards the window and to show that she is aware of the observer, and might thus keep her gaze straight ahead on the path. Likewise, the person inside will try to disappear behind the panes, through keeping on the side of the window, disappearing behind the curtain, or through peering outside from an angle (or metaphorically disappearing 'behind' the baby on his or her arm). Here, both parties are aware of each other through sideways glances from the corners of their eyes, which, in contrast to using binoculars, are sometimes fully reciprocated. In this case, instead of eye-to-eye contact and focal vision, the mutual awareness of self and other through peripheral vision is the more sociable kind of vision.

Similar considerations apply to situations of boat travel, and communication between leaders and followers, as described earlier in this book. When using the binoculars out on the sea or in the fjords, vision is not reciprocated, whereas when noticing each other from a distance or whilst driving along or following each other, the two parties usually do not look at each other in a focal kind of way, but are nevertheless mutually aware of each other's presence through the corners of their eyes. We could conclude, then, the closer two persons come to each other, the more they communicate through sideways glances, through the peripheral field of vision. This conclusion even applies to human-animal relations: when the hunter scans for seals through his binoculars, the seals do not look back and are unaware of the hunter's gaze. But in the more close-up engagement of the hunt itself, both hunter and animal are aware of each other's presence and movement through peripheral vision without looking at one another directly.

This thesis of the sociability of peripheral vision applies not only to East Greenland, but also seems much more broad. For example, Vergunst (formerly Lee) and Ingold's research on walking in the city of Aberdeen in northeast Scotland has shown that when companions walk side by side, they are visually aware of each other's presence, even though both are looking directly ahead (Lee and Ingold 2006; cf. Ingold and Vergunst 2008). Sharing the same visual field, communicating through the postures in movement, and shared rhythms and bodily engagements

are crucial to the sociability of walking together. Much has been written about
the sociality of vision, which purports to show that the more eye-to-eye contact
there is, the more interest and involvement exists between the participants (Hargie
and Dickson 2004: 107; cf. Argyle and Dean 1965; Simmel 1969). In an essay
on the sociology of the senses, Georg Simmel argued that eye-to-eye contact
'represents the most perfect reciprocity in the entire field of human relationship',
which induces a 'union' between the persons involved (1969: 146). This union,
he continues, 'can only be maintained by the shortest and straightest line between
the eyes' (*ibid*: 146). My ethnographic material from East Greenland, however,
suggests a radical revision of this idea. As I have illustrated above, and will show
with further examples over the course of this book, many situations of interpersonal communication in East Greenland are characterised by the avoidance of
directly looking *at* each other, or by averting gaze when doing so. It is through
looking *with* each other, i.e. sharing the same visual field, and coordinating one's
perception with another person's perception through the mutual responsiveness of
peripheral vision to each other's movements that vision becomes social.

Let me return to the theme of curiosity and the crannies of a house. In sum, we
are dealing with a dialectical process. On the one hand, seeing and listening to
what is happening on the outside or inside is channelled through the crannies of
the house. The house, with its windows and doors, structures communication in
that it lets inhabitants perceive only some parts of what is happening outside or
inside, depending on the position and size of the windows (and doors), the location of the house (since from some houses you have a better view to the harbour
or village centre), and so forth. On the other hand, inhabitants use these crannies
in particular ways: they are used to garner information on other people or happenings outside, which only in some cases and situations may be communicated
to others. Likewise, there are particular rules on when, what, and whom you may
perceive when approaching a house from the outside. This dialectic, nevertheless, has been undergoing changes with the transition from turf and earth houses
to modern Scandinavian house types. The multi-family turf houses common in
East Greenland until the beginning of the 20th century entailed the sharing of
domestic space with several other nuclear families and allowed much less privacy
than today. Nevertheless, it is probable that rules about not paying attention to
things happening very close by were followed much as they are today. Hence,
the fact that nuclear families started to inhabit their own houses has negatively
affected what people *can* actually perceive and know about their neighbours. The
change from multi-family dwellings to smaller turf houses preceded the introduction of Scandinavian-type houses only by a few decades. Nevertheless – through
their physical features – the newer houses afford a greater 'permeability' and an
increase in inhabitants' abilities to see and perceive people on the outside, which
again impacted on their communicative practices. The boundaries of houses have
changed, with bigger and more numerous windows, doors, and so forth. Turf
houses, by contrast, had only small windows made from skin that admitted light
through the front wall. Due to size, materials, and location, the houses afforded
much less visibility to the outside than their modern counterparts. This shows,

as Therrien has observed, that 'the modern home just as other non-traditional technologies entails new relations concerning the self, the other, and the exterior' (1987: 45, my translation). One could say that houses have been provided with 'eyes' that open up possibilities to gather information about life outside. The performance of those 'eyes' has been improved through the introduction of modern technical equipment such as binoculars. Communicative practices regarding other people's privacy have adapted to this greater 'permeability' in that particular rules have become established concerning when and in which situations a person may or may not pay attention to what she is able to see through the crannies of the house, often by way of peripheral vision.

Conclusion

People's communicative practices are bound up in a complex interplay between the kinds of spaces in which they take place (e.g., in the house) and the specific people who inhabit them. Family life within domestic spaces and communication among cohabitants have peculiar characteristics relating to the familiarity and intimacy of the people in question and are aligned to the material features of this setting. At the same time, characteristics of the communicational processes I observed are certain general features that come to the fore in a variety of situations and settings of the Iivit's lifeworld. These features include the high value placed on personal autonomy in family members' and cohabitants' communicative encounters, and the underlying tension between respecting other people's autonomy and the various social obligations that frame family life and community cohesion more broadly. This chapter has shown that houses are both performed and used in particular ways, and are to some extent adapted to by inhabitants. Whilst, on the one hand, communication at home draws attention to certain core values and ways of communicating in East Greenland, on the other hand they are not only 'used' for communicating in particular ways but may also alter communication. For instance at home, spoken communication is somewhat more important than when travelling on the sea or in the fjords, which is influenced by the comparably low level of background noise and the spatial proximity of the people involved. Moreover, the fact that people are wearing fewer clothes in houses with relatively high room temperatures allows for different forms of bodily communication compared to when being outside. For instance, limbs are more accessible to be used as 'illustrative material', and to carry out particular body movements for communicative purposes including play of various sorts. The body and its features are more visible, which influences its role in providing a context or incentive to be talked or joked about. Likewise, the 'warmth' of the house supports and makes possible various forms of bodily involvement such as sitting close to one another, hugging, sexual encounters, and so forth.

Moreover, it is important to acknowledge that houses have histories, and that people's interactions with, and uses of, space parallel the ways dwellings have been used in former times. The transition from turf houses to modern buildings has influenced and altered communication patterns, as for example windows with glass panes influence people's abilities to look out and observe others, and the introduction of separate rooms makes possible some sort of spatial privacy. The material objects with which contemporary houses are equipped, such as television, photos on the walls, and so forth, have altered communication patterns, not only in eliciting particular themes of conversation, but also through affording different kinds of communication. For example, the television, which is frequently commented upon, and similarly the radio have most certainly led to a reduction in storytelling that was so widespread in former times, a theme which among others will be addressed in the next chapter.

Notes

1 A figure by Petersen shows an average number of 19.3 inhabitants in 1928, which decreases towards an average number of 8.9 in 1940 (2003: 145).
2 I am mainly talking about the one-family homes that prevail in Sermiligaaq, other villages in the district, and Tasiilaq. My discussion does not include the bigger apartment blocks and newer kinds of multi-family buildings that you find in Tasiilaq and other urban areas of Greenland.
3 Men's tasks within the house often involve taking care of the oven, re-filling the water supplies, fixing work utensils, or pursuing other job-related activities. I rarely experienced men taking over the cooking or other types of housework.
4 For similar observations from other Inuit communities around the circumpolar North, see Briggs (1993); Goehring and Stager (1991); and Stern (2003).
5 Sermiligaaq does not feature satellite TV and there is only one TV channel available which is KNR-TV broadcasted by the Greenland National Broadcasting Company (*Kalaallit Nunaaata Radioa*). The TV programme is normally either in West Greenlandic with Danish subtitles or the other way around. Sometimes English films are shown with Danish subtitles, and once in a while you find programmes in another Scandinavian language.
6 Ivor Streck reports similarly from the Hamar in Southern Ethiopia, amongst whom knowledge is 'generally individualized and specific' (2004: 187). This goes along with a strong rejection of authority, at least according to the Hamar's rhetoric. In practice, nevertheless, he describes a tension between hierarchy and anarchy, equality and inequality, and individualism and sociality.
7 Kepa Korta (2008: 1658) writes that, 'Malinowski's considerations of language anticipate (though rather sketchily and without much theoretical analysis) many of the topics addressed in contemporary pragmatics'. Accordingly, Malinowski has sometimes been regarded as a predecessor to the linguistic work of Austin or Wittgenstein, though Korta's (2008) examination of the possibility of this influence finds no proof for this.
8 The term *sila* means everything outside of the house such as air, weather, or world; at the same time it stands for intelligence, mind, or temper.
9 For some further elaborations on the communicative functions of silence, see Basso (1970); Jaworski (1997); and Tannen and Saville-Troike (1985).
10 Breast-feeding did not happen very often in Sermiligaaq during my fieldwork. In the years 2005 to 2007 amongst other things because of the greater independence for a mother when feeding the child with a bottle, as some women have explained to me. The latter allows mothers to leave their babies behind at the grandparents' or other relatives' place and to go to work, attend school, go dancing, or spend some time at another settlement. During my more recent visits in East Greenland, nevertheless, the number of breast-feeding mothers I have encountered has substantially increased, and it seems to be much more common now.

11 For some further elaborations on communication and involvement from a primarily linguistic perspective, see Tannen (2007).

12 Erving Goffman has coined the term 'non-person' for individuals who are physically present but interactionally absent, at least temporarily (1959: 151–2).

13 I am aware that the distinction inside and outside is not universal and that one should not use it unquestioningly. I use it here as a local category since East Greenlanders frequently talk about inside or outside – for example, when walking in and out of the house and being asked where they are going. They have not explained to me where exactly the 'inside' of a house ends and the 'outside' begins, but from how they have used the distinction I assume that the doors and walls of the house are relevant in this respect. Therrien, moreover, has argued that the traditional Inuit habitat is symbolically based on opposition interior versus exterior, which is manifested linguistically (1987: 43).

5 Shared hospitality

Flows of guests, goods, and gifts

This chapter carries on the focus on the house but looks at it from a slightly different angle. I will take a closer look at sharing practices of different kinds, all in relation to processes of visiting. Sharing, which among Greenlanders is often considered a 'genuine Greenlandic cultural trait' (Nuttall 1992: 148), involves not only material flows, such as foodstuffs, equipment, gifts, or money, but also people: visitors and new household members, for example when adopting and fostering children, or when passing on names from deceased community members to newborn children. In the following chapter I will examine sharing in relation to visitors and overnight guests who enter and leave a house, and bring along or receive certain goods and services, yet I will not specifically deal with children who move between and among households. Sharing concerns a great variety of goods, both tangible and intangible. Elaborating upon sharing practices opens up questions of ownership and property relations, and calls for some theoretical grounding in the themes of reciprocity and exchange.

Property relations, reciprocity, and sharing

According to a common understanding, the concepts of reciprocity and sharing build upon a fundamental distinction between givers and receivers (MacDonald 2008). Givers are thought to hold particular rights over the goods they distribute, which are often discussed in terms of property relations (cf. Hann 1998b). In East Greenland, for a long time, the notion of property more or less coincided with the collective use of resources – with exceptions regarding a few personal items (Thalbitzer 1941: 625). Collective property relations still fundamentally shape the local economies of contemporary East Greenland, though the manifold changes since late 19th century colonisation have entailed more exclusive personal property rights with respect to various (mostly imported) goods. Nevertheless, and as I will illustrate below, a distinction between 'private' or individual and collective property rights is often rather misleading, since these categories often overlap and are not clearly separable (cf. Ingold 1986: 223). For instance, store-bought food or other imported items are sometimes shared out; however, not on a voluntary basis, but because of other people's entitlements towards these goods. This runs somewhat counter to a popular understanding of individual property.

In common Western academic parlance, property is not an inherent attribute of things or activities but lies in the rights particular people hold over things (Hann 1998b). Property may be considered as 'a network of social relations that governs the conduct of people with respect to the use and disposition of things' (Hoebel 1966: 424, cited in Hann 1998a: 4). 'Things', in this respect, are not confined to material items, but also construed more broadly to include intangibles such as names, reputation, stories, knowledge, and other non-material resources. Property is often regarded in terms of people's claims and the distribution of social entitlements (Hann 1998a). Nevertheless, following Barbara Bodenhorn, it is important to add that it entails not only claims, but sometimes also responsibilities for, or obligations with respect to, a certain good (2004: 39).

There have been persistent debates in anthropology around questions of exchange, reciprocity, and sharing, especially with respect to the analysis of hunter-gatherer economies. One important early contributor to the debate was Marshall Sahlins, with his famous essay 'On the Sociology of Primitive Exchange' (1972). He describes a continuum of reciprocity, from 'generalized' through 'balanced' to 'negative', which corresponds to a scale of kinship and social distance. As you move from family and residents towards strangers and enemies, the character of exchange shifts from the generalised towards the negative pole. Sharing is subsumed under generalised reciprocity, which is defined as giving without expectation of return. This type of reciprocity is said to prevail among family and household members. Balanced reciprocity is regarded as giving with the expectation that goods or services at equivalent value will be returned within a specific period of time. This type of reciprocity is less personal, and takes place in wider social sectors. Negative reciprocity is the unsocial extreme of the continuum, implying means of acquiring goods without the consent of the giver, such as theft. It is impersonal, involves opposing interests, and appears among enemies.

Although very influential in economic anthropology, this framework has been criticised on a number of grounds. Ingold, for example, refutes the correlation between social distance and reciprocity, based on the primary unit of the household (1986: ch. 9, 1999b). He argues that hunting and gathering societies are 'not internally differentiated by boundaries of segmentary exclusion into relatively close and relatively distant sectors' (1999b: 401). With family relations extending across the entire community, the internal cleavages are not between households but between genders and between generations. He demonstrates that generalised reciprocity, for example, may also occur between individuals who hardly know each other and whose relationship is characterised by social distance, as has been shown in various ethnographic accounts. However, following Ingold and a number of other scholars (e.g. Price 1975; Woodburn 1998), I argue that sharing should *not* be understood – in Sahlins's terms – as a species of reciprocity. Rather, I define sharing in terms of the activity of dividing into shares and allocating these shares to various recipients, without necessarily presupposing that 'stuff held at the outset by a single person is *divided up*, so as to be available for use by an aggregate of beneficiaries' (Ingold 1986: 233).

With respect to hunted and gathered foodstuffs in East Greenland, I concur with John Dowling's argument that 'the rights and prerogatives entailed in ownership are primarily those of performing the distribution, not of deciding whether or not the animal will be distributed' (1968: 505, cited in Ingold 1986: 227; cf. Boden-horn 2000b). In the East Greenlandic lifeworld, a hunter does not own the prey he has taken but appropriates it 'on behalf of the collectivity' (Ingold 1986: 227). Hence, as Ingold states, the 'possession of harvested produce, just like the pos-session of resource locales, turns out . . . to be a matter of custodianship' (*ibid*). Accordingly, one can draw a fundamental difference between 'sharing in' and 'sharing out'. Since access to and use of resources embraces the whole collectivity and is not confined to particular people, all 'sharing out' must always be under-written by the positive principle of 'sharing in' (*ibid*: 333–4). In the light of this, one has to reconsider the distinction between givers and receivers.

This has been formulated in rather similar terms by Nurit Bird-David (1990), who distinguishes between a 'giving' environment and a 'reciprocating' environ-ment. The former, in many ways, aptly describes the situation in East Greenland. Bird-David explains:

> This perspective suggests that gatherer-hunters are distinguished from other peoples by their particular views of the environment and of themselves and, in relation to this, by a particular type of economy that has not previously been recognized. They view their environment as giving, and their economic system is characterized by modes of distribution and property relations that are constructed in terms of giving, as within a family, rather than in terms of reciprocity between kin.
>
> (1990: 189)

This economic system 'implies that people have a strong ethic of sharing and at the same time practise demand sharing; they make demands on people to share more but not to produce more' (*ibid*: 195, on demand sharing see also Peterson 1993). Accordingly, in East Greenland shares may be demanded, as will be illus-trated below.

Bird-David's distinction builds upon a conceptual difference between sharing (or giving) and exchange or reciprocity. Reciprocity, as most famously proposed in Marcel Mauss's (1923-1924) treatment of gift exchange, is based on an unbal-anced symmetry presupposing incipient equality between givers and receivers. It requires a counter-gift, and creates a debt that has to be paid back. Sharing, on the contrary, is not based on the expectation of reciprocity. It is based on a fundamen-tal presupposition of equality between givers and receivers; it is often obligatory, and not primarily based on generosity (Ingold 2000: ch. 3 & 4; Peterson 1993; Price 1975; Woodburn 1998). James Woodburn explains this in his article, 'Shar-ing is not a form of exchange':

> Those who provide goods for such sharing or for such redistribution do so, often unenthusiastically, because they have to do so. They are dispossessed

of part of what is seen as theirs. They derive little personal advantage from such transactions. Those who receive the benefits regard them as entitlements which do not have to be earned or even acknowledged.

(1998: 63)

Thus, as opposed to exchange or reciprocity, sharing is less an economic transaction than an affirmation of autonomy and equality (MacDonald 2008: 7). People are obliged to give if they have, and economic goods and services are allocated without calculating returns. In order to explain the East Greenland case with respect to some of these broader theoretical issues, I will now set out and elaborate upon my ethnographic materials.

Sharing the home with visitors and guests

Members of most East Greenlandic households regularly share their living space with other inhabitants, and making visits and receiving visitors and guests forms an important part of social life. Whereas East Greenlanders use the term *pulaarteq* for both visitor and guest, I speak about a visitor with respect to short-term visits on a daily basis lasting from some minutes up to several hours, whereas the term guest will be used for a person who is accommodated overnight. In contrast, a more permanent resident of a settlement is identified according to his or her village (or town) affiliation, as for example Sermilingaarmeeq for an inhabitant of Sermiligaaq, or Tasiilarmeeq for a person currently living in Tasiilaq. These denotations usually change (depending on self-identification) when a person moves to another settlement on a more permanent basis.

Especially in winter time, many long evenings are spent together within family circles at people's homes. On these evenings there are frequently visitors around who are often – but not exclusively – members of the wider kin group. Apart from going to the meetinghouse, meeting outside in the village, and once in a while joining a special village event or festivity, visiting other people's homes is one of the major social activities in Sermiligaaq. Young unmarried people spend comparatively more of their spare time in the meeting house (*katersutarfik*) or outside with friends, whereas for middle-aged and older generations receiving or making visits at private homes are often more important. Visits usually include eating together, with the offering of food being an important part of hospitality, talking and telling stories, and sometimes watching TV/DVD. Larger family reunions are particularly important during the major celebrations of the year such as Christmas and Easter.

Apart from shorter stays on a daily basis, during many visits people stay overnight, which may last from some days up to several months or longer. One can observe much population movement throughout the region, for both several day visits and more permanent relocations (cf. Perrot 1975; Robert-Lamblin 1986: 47–52).[1] Movements between the villages and Tasiilaq are especially frequent, but Tunumeeq also move regularly between different villages, or decide to relocate to the Greenlandic west coast or Denmark. Motivations include educational and professional purposes, wanting to visit family or friends, a boy- or girlfriend living in another settlement, or merely striving for a change. Young and middle-aged single persons are especially active in this respect, as they are usually more mobile than others, and often visit and rejoin different households. Due to the flux of people who are accommodated for shorter and longer periods of time, the household has to be regarded as a fluid entity; guests may easily become (or cease to be) new household members. In the following section I will mainly talk about visits that take place within the Ammassalik region, and I will more or less set aside population movements between the east and west coast of Greenland, or Denmark and other countries.

Visitors are often relatives living nearby, people who pop in and out on a regular basis at any time of day. Visits also regularly take place among friends and village inhabitants more broadly. However, not every person who enters a house is regarded as a visitor, and a person who just enters without receiving hospitality, or contributing to it by him- or herself, is merely spoken of as somebody who has entered (*iserpoq*, to enter). Overnight guests, by contrast, are often related through kinship to household members.

Invitations, planning, and appointments

Visiting other people's homes is often spontaneous. Inhabitants may invite each other over without specifying a preferred day and time. *Pulaartetsauli/pulaartet-sausi* (you should visit us, sg./pl.) is an expression frequently heard. Likewise, a friend of mine once told me at the beginning of my fieldwork that '*maleq amma-voq*' (the door is always open). Thereby, hosts express their general interest in having a particular person over, and the invitee may choose him- or herself if he or she wants to follow and which day or time pleases him or her.

If a particular day or time of day is suggested, for example in response to somebody asking *qarungu?* (when?), answers are often relatively unspecific. One might suggest *arangi* (tomorrow), and then add the suffix *uppa* (maybe). *Arangi*, most often, is not taken literally as tomorrow but understood as 'not today, but I do not know when', 'at some point in the future', or – depending on the context – as a careful way of declining a request (cf. Gombay 2009, on use of the expression tomorrow among Canadian Inuit). If two persons agree on a particular date, these dates often constitute a vague agreement instead of a more or less fixed plan, with both parties taking into account that plans frequently change. During my time in East Greenland, it often happened that individuals failed to show up on a previously agreed day as they had come to prioritise other activities in the meantime. Such behaviour was not meant to cause offense to the other party. The person not following the invitation was usually not asked for an explanation, and questions were raised only sometimes. Nevertheless, if invitees failed to show up on a regular basis without reasonable explanation, the inviting party would draw their conclusions, which influenced the relationships at stake. Whilst personal decisions are not openly judged by the people around and it is more or less up to the individual to decide on his or her actions, he or she has to bear the consequences of his or her behaviour. Hence, some kind of judgment indeed takes place, though most often it is not expressed directly in the face of the person concerned, but manifests itself through people's practices and behaviour towards the person.

It is important to add that not all invitations are equally flexible, such as on special festivities, birthdays, and religious holidays. Such days are kept in mind and may hold a relatively high priority – though here, too, the above rules of not questioning, judging, or interfering apply when somebody does not attend. Moreover, specific invitations are articulated on a very short-term basis, and East Greenlanders regularly call or send messages to friends asking if they want to visit or to be visited at that particular moment, for example when inviting family members and friends to come over to have freshly caught meat or fish, to join a birthday party, or when asking others to come visit in order to play cards.

Nevertheless, Tunumeeq ask quite frequently about fixed dates concerning future plans, and visitors, for instance, are questioned about dates of their departure or return to the village. I was initially a little surprised that I was expected to answer these questions with a firm date (at least appearing 'firm' to me), despite the common knowledge that these dates often change, for example because of delayed helicopters. When I was questioned in this respect, I would mention a day but add a 'maybe' since I was not sure about the weather, possible transport, and the like. At this, some people would look at me with some astonishment, insinuating that I did not know what I wanted. On these occasions, lack of specificity was not appreciated. My surprise reflects the fact that I was accustomed to mentioning a date or time only when being relatively sure about it, whilst such certainty did not seem to be necessary for most Tunumeeq.[2]

Flexibility in ways of dealing with appointments and planned dates has frequently been mentioned in the literature on Inuit people around the circumpolar North (Briggs 1993; Goehring and Stager 1991; Stern 2003; Stevenson 2014). This is not because of any indecision or unreliability, as outsiders sometimes assume. Instead, it has to do with a different attitude towards future planning and being on time. One encounters a great readiness to re-evaluate plans according to current situations and affordances. A planned date *per se* is not unchangeable, and the wider contexts and wider range of available alternatives play a decisive role in its reckoning. Other inhabitants accept that a person is not able to, or does not want to, accept an invitation, which can merely relate to the fact that he or she was in a bad mood or tired, and probably nobody would have enjoyed a previously planned gathering. The non-interference in a person's decisions reflects the idea that one can neither predict, nor know another person's motivations, nor control others (cf. Morrow 1990: 149). My experiences thus support Nicole Gombay's argument that 'the conception of time developed by and inculcated in Inuit is essentially contingent, that is, it depends on a complex state of affairs which affect how, when, and why people act' (2009: 2). This contrasts with Euroamerican attitudes towards time, in which 'time is essentially perceived to be something that is absolute' (*ibid*).

On a broader level I believe that these attitudes can be explained by a general recognition of indeterminacy among the East Greenlandic Iivit. As Morrow (1990: 145) writes, 'In Euroamerican societies, we act as though we can discern definite meanings, and tend to assume that they are accessible to analysis'. In contrast, she argues that among Yupik people, 'meaning is treated as essentially indeterminate' (*ibid*), a statement that seems applicable to East Greenland as well (for similar accounts of Canadian Inuit, see Briggs 1991; Therrien 2008). I have encountered a great awareness of the changeability and instability of the world; things are not conceived of as bounded or as happening in discrete parcels. Just as Gombay remarked with respect to Canadian Inuit, Tunumeeq operate on the assumption that 'everything is interconnected, and actions can have numerous, often unforeseen, consequences' (2009: 3). This, again, shows that, on a fundamental level, Inuit are taught to think in the present and to accept the fact that plans change. Though Iivit did and do think about the past and make plans for the future, they do not 'operate under the assumption that the future is something that can be controlled' (*ibid*: 6–7; cf. Briggs 1993). Similar attitudes have been described for the Cree of Northeastern Canada, for whom, according to Harvey Feit (1996: 435), 'humans do not ultimately control life, but intimately and respectfully link their thought and action to those powerful beings who create the conjunctures of life'. Nuccio Mazzullo and Tim Ingold (2008) have found the same attitude among the Sámi in Finland.

Let me now turn to the actual practices of visiting. In order to explain the procedures of entering other people's homes, I will start with a short ethnographic anecdote from my fieldwork in 2007 (field diary, April 13, 2007).

Figure 5.1 Sermiligaaq, June 2005
(Photo: Sophie Elixhauser)

Entering a house and greeting and receiving guests

It was an afternoon in April in Sermiligaaq. I was helping a friend to hang the wet laundry from the clothesline spanning the veranda and the stairs leading to the entrance door of her house. A woman approached the place. I had not seen her before. She opened the gate at the foot of the stairs, and slowly ascended the steps towards us. She did not say anything. Likewise, my friend paid no attention to her, and I followed her example. The woman stood right next to us for some minutes, and I was somewhat surprised by her merely looking at us. My friend and I continued our business, and now and again we exchanged some words between the two of us. It was quite cold and our fingers were freezing. After a while, the woman entered the house. A little later, we finished our work and followed inside.

My friend's husband and their children were at home. The visitor had taken a seat on one of the chairs in the living room. She did not speak and was neither addressed nor spoken about. Initially, I noticed some surprised faces among the family members, but soon nobody paid her any attention. At one point, the woman got a cup from the kitchen, and poured herself some tea from the thermos flask in the middle of the dining table. She went on to help herself to a slice of bread from the kitchen. Some minutes later, my friend exchanged the first and only words with the woman – at least from what I heard – and offered her some seal meat. The woman seemed to be quite hungry. A little later the family members and I took our seats on the sofa, and we began watching a DVD. My friend's husband started to imitate a scene from the DVD, making some funny gestures and pretending to hit somebody in the back. While doing so, he looked over to the visitor who had turned her back towards us and could not see us. We had to smirk. It was very obvious that my friend's husband was making fun of the woman. I left the house some time later, and in the evening I saw the woman walking outside in the village. The next day, when I asked my friend about this incident, I was told that this woman is a person with disabilities from another settlement, visiting relatives in Sermiligaaq. She has to be pitied, my friend explained to me.

In some respects, this episode is rather unusual, above all because the woman hardly communicated with us during the several hours of her visit. I have met many East Greenlanders who are not too talkative. Sometimes minutes may pass during visits in which nobody actually speaks. Yet silence for such a long time, accompanied by very little social interaction, is rather unusual, especially considering that the woman was neither a family member nor a regular visitor. Most certainly, this behaviour related to the woman's disability, at least to some extent. When offered seal meat, she was treated with the obligatory friendliness which visitors usually receive, but, apart from that, my friend and her family did not pay the woman much attention. They did not look at her directly and she was hardly spoken to nor greatly acknowledged in other ways. Nevertheless, and in spite of these rather unusual characteristics, this story illustrates some very common communicational modalities and attitudes with respect to entering other people's houses and receiving visitors in East Greenland.

The inhabitants in the village may, relatively, freely enter a house, even if nobody is home (cf. Nuttall 1992: 114, reporting similar from rural West Greenland). People often enter without knocking or ringing a bell. The doors to most houses in Sermiligaaq are unlocked most of the time. Exceptions are some newer types of houses where the main door can only be opened from the inside. However, there are only few such houses in Sermiligaaq and most buildings can be opened from the outside. These houses are usually only locked when all household members have left the village for a longer period of time, i.e. at least overnight. During the day, the entrance doors are left open, even if temporarily nobody is home. Referring to the Canadian Arctic, Therrien (1987: 44) sees in this a continuity with respect to the use of space in snow houses from former times, which may be extended to turf houses in East Greenland.

Upon entering, visitors usually pass through the outer porch (or the two entrance rooms, depending on the type of house), and open the door to the main living space. They might walk inside and not say much, or utter a short greeting. At other times, visitors stand in the doorway and ask or call for a particular person. If they are given no reply, they might take off their shoes and look inside the living room and kitchen. Whereas living rooms and kitchens may be entered freely and are frequented by a variety of people, bedrooms are not accessed without the approval of one of the household members. Visitors also sometimes announce 'I want to visit' or 'I am here in order to visit' (*pulaarngusuujarngitsi/pulaarartaqqalerpua*) upon entering the house, which is positively affirmed by the household member who feels addressed and might answer *iserniaat!* (come in, sg.)

Walking in and out of somebody's home is only sometimes accompanied by verbal greetings. Indeed, the majority of such expressions have been introduced rather recently. When Holm visited the Ammassalik region at the end of the 19th century, the inhabitants did not use specific words to express welcome, though he encountered some common forms of saying goodbye, for example 'take care of yourselves' (1911: 141). Nowadays, regular ways of saying hello are *kulaa* (from dan. *goddag*, have a good day), *aluu* (hello), or silently raising eyebrows, opening the eyes, and looking at somebody. Shaking hands is quite common with people one has not seen for a while. Close family members, who are welcomed back after some period of time, are hugged and kissed or sniffed with the nose. Goodbye, most generally, may be communicated through saying *bye*, *tagiu* (see you, or literally 'we see each other'), or *kunnat* in the evenings (from dan. *godnat*, good night).[3] All the same, Tunumeeq often arrive at or leave houses without communicating a word of greeting or farewell. This is more frequent in the case with regular guests and visitors such as neighbours, members of the wider family living close by, as well as children. Considering the many expressions with a Danish origin, such as *kulaa* or *aluu*, the effect of Danish in increasing the prevalence of verbal greetings is quite apparent. In the presence of Danish residents and foreigners, Iivit much more often say hello or goodbye, compared to when they communicate amongst each other. I, too, was more frequently greeted than other people in the village (or town), though this somewhat lessened as my fieldwork progressed.

Every community member may enter any house without prior invitation or request. Reasons may range from wanting to warm up, looking for company, or searching for a particular person, to a variety of other motivations. My friend Maline explained this to me at one point on a winter's day in 2007. We were standing next to the heliport waiting for some family members who were supposed to arrive with the helicopter. It was cold, and I mentioned to Maline that I was freezing. She suggested that I enter the house next to us in order to warm up. I was somewhat reluctant to do so since I had not been invited to that place before and did not know the people well. But Maline explained to me that I was free to enter any house. Hospitality is such that people's requests to enter are rarely refused, except if there is a reasonable explanation.[4] That is why the disabled woman was allowed to walk into my friends' house without being invited or being particularly welcome. Passing the entrance door was decisive in this respect, since joining somebody on the veranda or somewhere in front of the house is not regarded as visiting. Nobody would have prevented the woman from entering, and it was up to her to stay with us if she liked. Moreover, visitors usually determine the length of a visit, though hosts might themselves depart the house and leave the visitor behind. They usually do not feel obliged to stop their current work or change their daily rhythm in order to accommodate visitors and guests, especially when visits come unannounced.

For instance, one day when I was about to leave East Greenland at the end of my year of fieldwork, I went over to a friend's house in Tasiilaq to say goodbye. When hearing about my departure the day before, she had told me to pass by without specifying a time. When I arrived at her house in the afternoon, she was tired. She offered me some food, and after I had started to eat she told me that she wanted to lie down and take a rest. She suggested that I stay and finish eating, and then take a seat in the living room. We said goodbye rather briefly, and she went to bed. I finished my plate, and left her house. My friend had not felt obliged to stay awake in order to take advantage of my last visit. However, she did show her hospitality by offering me food and suggesting that I stay afterwards (her children would have kept me company). The point is that she was not obliged to pretend to enjoy socialising whilst actually not feeling like it. Also, returning to the earlier example of the disabled woman, my friends did not greatly bother to entertain the visitor. I have experienced many similar situations in which I had passed by some friends' houses, and a particular friend was just about to take a shower, to pick up a child from the kindergarten, or to leave for the shop. I was either told to accompany that friend, or to wait until she was finished or had returned. Sometimes some children were asked to keep me company so that I did not have to wait by myself. When hosts are busy at home, visitors sometimes stay around and give a hand, just watch, or entertain their hosts. Other villagers' daily rhythms, however, are usually well known by residents, which allows for the times of visits to be adjusted accordingly. Moreover, the presence of overnight guests, for shorter or longer periods of time, does not usually imply that hosts spend much additional effort taking care of them, through preparing extra dishes, adapting daily routines, and the like. This may sometimes happen but is not necessarily expected. All in all, I found that most often the presence of visitors is remarked upon very positively, even when hosts do not have the time to take care of them extensively. Unwanted visitors, in contrast, are ignored.

Thus, on the whole, visitors in East Greenland are treated with courtesy and respect. This obligatory politeness has been mentioned by several ethnographers who have worked in the region (e.g., Holm 1911: 141; Gessain 1969: 112; Robert-Lamblin 1986: 145). A visitor may join the inhabitants of a house, help him- or herself to tea, water, and coffee or bread, and is usually offered food. Hospitality seems to be based upon a different treatment of drinks and bread, which is available in most houses most of the time, on the one hand, and warm meals – Greenlandic country and Danish dishes – on the other hand. Drinks and bread may be taken by a visitor or guest without being invited, whereas eating a warm meal (or any cold meal apart from bread) presupposes an invitation by the hosts. Such obligatory courtesy, moreover, also concerns matters of accommodation, as will be explained in the following section. In order to illustrate some of my arguments, I will start with two stories from my fieldwork in 2007. The first episode closely follows my field notes recorded in Sermiligaaq, April 10, 2007.

Accommodating guests: negotiations, obligations, politeness

As so often happened, I was spending a peaceful afternoon at Sibora and Bianco's place. My elderly friend Sibora was home alone and she was happy for some company. After a while her husband, Bianco, entered the house. From his and Sibora's talk, I picked up that it was their son Vittus's birthday. Vittus lives with his family next door. I was surprised that nobody had been talking about this event before, neither Sibora nor Vittus and his wife Dorthea, whom I had met earlier that day. Bianco and Sibora were searching for their present, and we headed over to Vittus's house. We were called to eat roast pork and potatoes with brown gravy. Freshly cut seal meat was cooking in a big pan of salty water, which we were offered later on. Shortly after our arrival, a male visitor entered the house. I had not seen him before. He was welcomed by our hosts and the other people present. He seemed to pass by quite spontaneously and did not know about Vittus's birthday. Seeing me among the group, he looked a bit surprised. Yet, initially, he did not talk to or ask about me. After we had finished eating, we sat on the sofa and drank coffee and tea. Listening to the conversations, I came to understand that the man had arrived with the dog sled from the neighbouring village Kulusuk, guiding a group of tourists. My companions cracked a few jokes, and suggested possible love affairs. The assumed 'very rich foreigners' might be a good catch, and everyone agreed. After some time had passed and news was exchanged, the man started asking about me. Might I be a new Danish teacher? He did not address this question to me but looked at the other people in the group. Sibora explained some words about my presence and that I was part of their group (*'taanna uangi'*, 'she belongs to us'). She answered his follow-up questions on my age and if I was here by myself. I listened to what was being said about me; I was not directly addressed. After we had enjoyed some of the cooked seal intestines, Dorthea asked the visitor where he was going to sleep that night. He said *'nalivarnga'* (I don't know). Dorthea continued asking: *'Tattanni ilangatsaulit?'* (Will you sleep here?) The Kulusummeeq answered cautiously that he still had plans to visit some other houses that evening, and that he did not yet know where he would end up staying the night.

Contrary to this story, in which a visitor from another village came to Sermiligaaq, the second episode I am going to present is about how I was part of a group of Sermilingaarmeeq who visited another settlement in the district. The story is based on my field notes of 8 to 11 June 2007.

It was June 2007, the narwhal hunting season. A lot of Sermiligaaq hunters were leaving for the Sermilik Fjord, located a day's boat trip from the village. My friend Hansigne, her husband, Mathias, and their three-year-old daughter, Dina, were also getting ready for departure. Mathias's two brothers joined the group, and so did I. Hansigne, her daughter, and I were meant to be dropped off in Tiilerilaaq, around halfway to the hunting areas. My friend wanted to visit relatives, and she was keen on attending the Bingo evenings. When we reached Tiilerilaaq, Hansigne, Dina, and I were let off at the dock. Whilst climbing out of the boat, one of Mathias's brothers asked me jokingly where I would be heading to in the village. I said that I did not know. When I had asked Hansigne about our accommodation before our departure, she had only vaguely mentioned that her father has relatives in Tiilerilaaq. The men continued their boat trip towards the inner part of the fjord. They were going to fetch us upon their return.

Carrying our luggage, we walked through the centre of the village. Not many people were outside. Hansigne tried to remember her father's relatives' names, and explained that an elderly woman living by herself would be the most suitable place for us to stay. She had a particular woman in mind, but was not sure about the location of her house. Next to the service house, we met a friend of mine whom I had come to know during some previous visits; she invited us over for a meal one of these days and mentioned that she would have liked us to stay at her place but that she did not have any room due to other guests. Asking her about the whereabouts of Hansigne's relative's house, she directed us to one of the old cabin houses at the other side of the village. The woman was home and invited us to come in. One of her first questions was: *'Sumo ilanngaatsausi?'* (Where will you sleep?) Hansigne said that we did not know. It took only a few words for agreement to be reached on accommodation at her place. We established ourselves and stored the groceries that we had brought along in one of our host's cupboards. After a while, the woman's son entered the house and brought a share of seal meat. We were invited for dinner. In the evening we headed over to an old villager's home where the daily gambling meetings took place. The house was packed with Tiilerilaarmeeq, and Hansigne met some friends and people she knew. Not everybody knew Hansigne though, and I heard some villagers explaining to one another that this is Sibora's daughter. I also heard curious inquiries about me from Tiilerilaarmeeq, addressed to my companions. Only a few people put such questions directly to me. Too much open curiosity is not appropriate.

After a night at our host's cabin, we encountered some difficulties. Hansigne's daughter, Dina, was used to moving around freely and accidentally disturbed some pieces of our host's living room decorations. Though our host did not explicitly say so, we noticed that she was bothered (Hansigne was aware of this earlier than I). The same afternoon we visited another elderly woman who was also related to Hansigne, and who expressed her disappointment that we had not chosen her place as accommodation. Her house was much more spacious, and we quickly decided to move. When we went to fetch our luggage from our first host's place, Hansigne merely declared that we were moving house, and the lady did not ask about our reasons. We removed our groceries from her cupboard in order to store them at the new place. Our new host cooked some seal meat, but the share she had received was very small, and she apologised for not being able to invite us to eat. We consumed some of our own groceries and enjoyed a relaxed afternoon at the new place. Later, we visited some other people's homes, and once again called in on the bingo place. After that night, however, we had to move house again. We slept late into the morning, and our second host made us understand that she was not pleased by that. Subsequently, she mentioned that she would have a home helper from the community coming to her house early the next morning, and that we would have to get up and leave the house beforehand.[5] Hansigne and I understood this as a subtle request to look for a new accommodation. Having joined in village life, Hansigne had met many people familiar to her in the meantime, and without much trouble we moved to another family's home. We felt most welcome at this place and stayed there until the hunters' return.

Most East Greenlanders very much enjoy travelling to other settlements in the region to visit friends and relatives, or sometimes in order to move more permanently. These trips are more frequent during the boating season from spring to autumn, as transport is cheaper. Accordingly, I often found that inhabitants from other settlements came to visit in Sermiligaaq for a longer period of time, or Sermilingaarmeeq left for Tasiilaq or one of the other villages. With respect to people arriving, decisions about whether to stay more permanently or to leave after some days, weeks, or months were often not reached in advance, and questions were answered by saying '*nalivarnga*' (I don't know); whenever plans were mentioned they were repeatedly modified. For example, my friend Hansigne did not know in advance how long we would be staying in Tiilerilaaq, though upon asking she proclaimed that she expected it to be 'some days'. In this case, it depended upon the availability of narwhal. This flexibility may have its roots in the semi-nomadic lifestyles of former times, when survival was dependent on taking quick decisions about when to move and where to settle.[6]

Greenlanders arriving at a settlement are usually temporarily housed by relatives, and most often by their closest relatives at the particular settlement. In a small region like Ammassalik, many inhabitants have relatives in other settlements, if only distant ones (and if they do not, they hardly travel to that place). Sometimes East Greenlanders also stay at friends' places, though this is more rare. Matters of accommodation are not always previously arranged, and if a person shows up, relatives (or also friends) are obliged to receive him or her. Of course, the availability of space also plays a role and is discussed, especially with respect to longer-term accommodation; yet I have hardly encountered a host who would openly refuse a request for accommodation, though in some cases alternatives may be cautiously suggested. For instance, a host might remark that there is more space at another house. Pointing out this possibility, nevertheless, would still leave the decision on where to sleep to the visitor.

I sometimes heard Tunumeeq discussing potential places to stay before leaving for a trip, which did not necessarily mean, however, that prior arrangements had been made with the potential hosts. Nonetheless, in more recent times, with landline phones connecting most households in the region and the enlargement of the mobile phone network to cover all Greenlandic settlements since 2006, it has become quite common to phone in advance and ask if accommodation is possible, and to make sure that during the time of visit the potential hosts will be at home. When calling a possible host in advance and asking for accommodation, or similarly when sending a text message via a mobile phone, intentions and wishes are often communicated somewhat more explicitly. These new technologies thus play a part in the development of a new etiquette with respect to visiting and verbal communication respectively.

When arriving unexpectedly at someone's house, at the outset, requests for accommodation are usually not directly addressed, but instead visitors might explain their plans and then wait for an invitation. Invitations are likewise framed in rather open ways, an etiquette which allows both visitors and potential hosts to mention different plans without having to confront the other party with a refusal. For instance, during my trip with Hansigne, the question 'Where will you sleep?' asked upon our arrival at our first host's house showed the woman's general interest in having us stay at her place, as similarly did Dorthea's question to the male visitor who had come from Kulusuk with the group of tourists. Since, in the latter story, the male visitor did not yet have concrete plans, the question was followed up with a somewhat more explicit invitation on the part of Dorthea, which the man met with carefully formulated vagueness.

Open-ended questions, for example about where visitors will sleep, can themselves be regarded as invitations, if they are not preceded or followed by a reason for not being able to provide accommodation. In my experience, explanations for not being able to host somebody would be mentioned immediately in order to prevent having to refuse a request, just as did the friend, for example, whom we met on the street on our first day in Tiilerilaaq. Without our even asking her about it, she explained that she would like us to stay over but that she did not have any room to accommodate us. From time to time during the initial months of my fieldwork I became confused by these practices, since I had not yet understood that questions about my plans were often meant to be invitations. A few times I waited for a more explicit invitation (or the repetition of an earlier brief remark), and was not quite sure if I was welcome at somebody's house. However, an invitation, in the form that I was used to, often did not follow. Framed as a short remark or question about my plans, the general willingness on the part of the hosts had already been clearly communicated, though it was not yet clear to me.

The example of our visit in Tiilerilaaq also shows that it is not appropriate to withdraw a previous offer explicitly, except when there is a reason that is deemed legitimate by both parties. Acceptable reasons may include, for instance: being short of space due to other visitors, or being away from the settlement for some time. Conversely, changes of mind and personal moods as well as aversions and interpersonal tensions are rarely verbalised explicitly, but instead are communicated subtly through the degree of attention given to a person, various clues, particular ways of talking and not-talking, humour, gaze, and so forth. Such practices are also to be found in Ittoqqortoormiit, way up north on the Greenlandic east coast. A Danish resident, who works at the weather station and has lived in the region for some decades, explained to me that when asking an inhabitant about being accommodated they would never decline a person's request, be it relative or stranger. The travel guidebook Lonely Planet once stated that a tourist could ask every hunter for accommodation. My informant was quite angry about this statement, which took advantage of local rules of politeness, as even hosts that would not be happy about such visitors could not refuse without losing face.

These subtle ways of communication are also evident in the story of the trip to Tiilerilaaq. Our first host's reactions to my friend's daughter's liveliness and our second host's critique of our sleeping habits, as well as her request that we should get up early the next day, were understood as hints that we should find other accommodation. Due to codes of politeness, neither of them could suggest that we should leave. Likewise, when we had decided to move and told them about it, they did not ask for the reasons, and thus bypassed a possible confrontational exchange. With respect to the second accommodation, we found out later that right after we had left the elderly woman had started to drink alcohol, and was either drunk or trying to recover from drunkenness during the days that followed. When Hansigne had discussed these issues with some Tiilerilaaq relatives, the latter had remarked that the explanation about the home helper had sounded like an excuse, since inhabitants are not usually asked to leave a house during these activities. They agreed that our host's plans to drink alcohol must have been her main incentive for wanting us to move out. Though a high level of alcohol consumption is common for a proportion of residents, it is frowned upon by many others and would not have been appropriate in the presence of non-drinking guests. Similarly, if potential hosts are drunk upon the arrival of guests, this is often taken as a reason to look elsewhere for accommodation.

Another observation that caught my attention in both of these stories as well as on many other occasions is that when a visitor arrives at a place and some of the inhabitants are not familiar with this person, it is not appropriate to ask the newcomer for identification, at least not right away. It was quite interesting to observe how the Tiilerilaarmeeq received Hansigne, a person from another village. Not everybody knew who she was, and not even all of her (distant) relatives recognised her at first sight. Thus, as with volunteering information to other household members, most villagers did not approach Hansigne to enquire who she was, but rather asked other people about her (much the same happened with me). Subsequently, a little genealogical information and mentioning that she was Sibora's daughter would be enough. The male visitor from Kulusuk in the story recounted earlier in this chapter, who was apparently quite curious to know who I was, did not address me and waited for some time until he started to ask my friends. Then, in my presence, information was exchanged about me, though not once was I personally asked to contribute to the conversation. The man assumed that somebody else would volunteer information about me. These practices illustrate that, on a broader level, asking unfamiliar people to identify themselves is not acceptable, though it is quite common to make a suggestion, such as 'Is she a new teacher?' Other individuals (who are familiar with the person), or the person herself, might then offer information. This indirect questioning, which I found to be common among Tunumeeq, underlines the importance of individual autonomy with regard to communication.

The examples I have provided above illustrate how situations are managed when a visitor's projected length of stay is short (a few days). Yet similar rules on hosts' and guests' obligations apply with respect to longer stays. People arriving at a new settlement will usually go to live at a close relative's home. If there are different households to choose from, decisions on where to stay are based on such factors as the availability of space and the number of household occupants. These decisions about where to stay are taken by the visitors themselves rather than the hosts; although the latter are more or less obliged to share their homes, the former are expected to take care in choosing a house where they are welcome and where there is enough space. If hospitality is demanded, hosts have little power to decline, due to the imperatives of demand sharing, as explained above. Nevertheless, demands are often communicated in a subtle way, in order not to impose wishes on others and to create obligations. Hosts and visitors or guests are attentive to subtle points of tension that may not be directly expressed because of the people's expectations of hospitality and politeness. Invitations, requests, and refusals are often 'hidden' behind the words, and it is the responsibility of the listener to understand and to decide. Kulick and Strout have observed similar communicative patterns in Gapun, a Papua New Guinean village. They record how a 'recipient cannot claim at a later point to have been "pushed" into doing something he did not want to do . . . The listener should feel moved to help the speaker out of his or her own sense of solidarity and goodwill' (1990: 297). Thus in East Greenland, as in Gapun, elaborate communicative codes reinforce the fundamental importance of the notion of personal autonomy.

On the subject of sharing a house with visitors and guests, it is important to recall that in the past, earth houses in East Greenland were communal property. Contemporary attitudes towards sharing houses (albeit the modern variants) still reflect these property relations in some ways; for example, attachment to particular buildings is often not very strong. Though the occupants of a house hold some privileged rights with respect to the buildings they live in, anybody may enter, take a seat, make use of the living space (at least the living room and kitchen), and be fed and accommodated upon request. Thus, building upon people's former semi-nomadic way of life, ownership of a house could better be framed in terms of what Ingold has described as 'custodianship', a fluctuating form of ownership in which, for an individual, a particular good or object 'may be exclusively his to dispose of, but it is not his alone to consume' (1986: 227). Up to now I have mainly described communicative exchange in relation to who may enter a house and visit. I want to focus now on the visits themselves, starting with what I see as an important motive for visiting: the sharing of news, stories, and entertainment.

Storytelling

The presence of visitors and guests at people's homes provides much occasion for conviviality. Communication during such get-togethers is often characterised by comments, jokes, stories, and comfortable moments of silence. Village and family news is exchanged, narratives are told, and people gossip and engage in different sorts of verbal and nonverbal play. A person may have much to say immediately upon entering, or remain silent for a longer while before starting to talk. Visitors are not usually confronted with a firm expectation of having to provide particular information, and for the most part, the choice of whether to speak or remain silent lies with the individual (cf. Morrow 1996: 414). There are exceptions, nevertheless, and sometimes visitors are asked to provide some sort of entertainment, e.g., through being told '*oralivarniaat!*' (tell us something!). Yet such a request leaves the choice of the particular topic and the way of speaking to the visitor. Sometimes, of course, specific questions are posed and particular themes are suggested. Such questions, nonetheless, are usually framed in a way which allows visitors to choose how to respond or not to respond at all.

One form of entertainment for which visits provide occasion is storytelling. I was often told that this practice was more pronounced in former times and has nowadays been partly replaced by modern media such as television, radio, computer games, and the Internet. Stories (*oralittuaq*), nevertheless, are still very much appreciated and form part of many get-togethers at peoples' homes.[7] Stories are told in a variety of settings, such as during social gatherings at community buildings or when meeting outside. People's homes, however, are the most common locations. Good storytelling abilities (*oralivattaqqippoq*) are particularly widespread among the elders. The themes and content of stories vary widely. Some report everyday news of trips, village events, and people. Others touch on the past and tap into themes that have been retold for centuries (for collections of East Greenlandic tales and myths, see Holm and Petersen 1912 [1887]; Rasmussen 1921). Scary stories (spine-chillers) are particularly liked (cf. Pedersen 2009a, 2009b). Storytelling performances by Greenlanders from other communities are now and again broadcasted by KNR (the national television channel) and are also sometimes available on DVD. These are much enjoyed by Tunumeeq.

The decrease in storytelling in East Greenland has also been reported by Robert-Lamblin, who nevertheless remarks on the persistence of hunting stories. She writes:

> The traditional hunter's narrative, with abundant detail and illustrative gestures, told when he returns home to his assembled family, is still performed by those who live by hunting. But fishermen and wage-earners do not have experiences worth recalling or that can win the attention and admiration of everyone when they return home.
>
> (1986: 135)

Likewise, I have more often experienced storytelling performances in hunting families. Journeys are often talked about, and recounting hunting trips is especially popular. Hunting stories that I was told often included a variety of sounds, hand and arm gestures, bodily movements, and facial expressions. This reflects what Holm (1911) observed at the end of the 19th century. He wrote that 'In these tales gesticulation, shrieking, and changes of voice are often of more account than the words and the actual run of the story' (*ibid*: 125). Gessain describes East Greenlanders' illustrations of hunting sequences through their finger games as a 'true gestural language' (1984: 88). In another work, he states: 'These people are perhaps the best storytellers in the world; with them storytelling becomes a theatrical spectacle' (Gessain 1969: 180–1). Another example of gestures whilst telling hunting stories is the use of the fingers to enumerate the number of prey taken (cf. Thalbitzer 1921: 148; Nuttall 1991: 43). Hunters I spoke with were often very keen to show me, with their fingers, the number of the animals they had caught in their lives. I frequently provoked these manual accounts by posing a question about a specific trip or type of hunting in former days. My questions often came from looking at a picture on the wall showing a hunting scene, or from being shown jewellery made from polar bear or seal. Pictures with photos are often talked about, referred to, or pointed at when telling stories about past happenings, and they give an incentive to speak or ask about former trips, experiences, or particular persons. Likewise, they also play a role in remembering deceased family members.

In this respect, I noticed that East Greenlanders I met would not usually start talking about personal success and accomplishment without being asked first. Showing off or trying to appear better than others is not accepted in East Greenland. In parallel to my observations, a Danish resident in Ittoqqortoormiit, married to a Greenlandic woman, remarked that while catches like polar bears or narwhals, for instance, contribute to a hunter's reputation, exact numbers would not be specified to others without being asked for them. One does not find open competition among the hunters; rather, somebody catching more than others is called 'lucky', while others are referred to as 'unlucky'. Nevertheless, I found that now and again some individuals seemed to appreciate being given an opportunity to talk about their personal performances (though not all, depending on the personality), and invited this line of questioning by, for example, showing me hunting trophies or products. Stories provide an opportunity at least to mention personal accomplishments, though these are not usually greatly elaborated upon.

A broader function of stories is that they 'prompt audiences to all sorts of conclusions without the speaker having to state them' (Brison 1992: 20; cf. Cruikshank 1998). They not only make it possible to highlight people's experiences whilst at the same time providing entertainment, but they are also a means to pass critical judgement and to point out values (cf. Fienup-Riordan 1994, 2005, on storytelling among the Yupiit). By eliciting stories, I was sometimes able to learn subtly about shared expectations and about my own infractions without being taken to task. For instance, in Chapter Three I mentioned the example of my having laughed in a dangerous situation during a hunting trip since I was not aware of the seriousness of the situation. Later on friends of mine who performed the story of our trip when back home in Sermiligaaq repeatedly mimicked my reaction. I have experienced many similar cases in which other village inhabitants and I were the butt of stories told at our expense. Through this, disapproved behaviour was not explicitly judged, but intonation, facial expressions, and gaze would clearly communicate the storyteller's opinion. The audience played a major role in this process as such stories were usually repeated several times, often in the presence of visitors and guests. Stories expressing particular values are also very relevant for the education of children, and have similar functions and characteristics as other non-confrontational communicative strategies described above, such as teasing and open-ended questions (as opposed to orders).[8]

Hence, above all, storytelling brings about a dissemination of individual experience and knowledge, not only with respect to people featuring in the stories and particular values, but also regarding particular hunting knowledge, about routes, ice conditions, or abundance of prey, among other things. As Heonik Kwon has stated for the Orochon in Siberia, Tunumeeq, too, 'tell stories, in which the animals' behaviour and their movements and the narrators' state of mind and actions are all represented in close connection to one another and to the landscape' (1998: 118). In this, the successful catch is not very relevant, and is often not even mentioned, again showing the people's general reluctance to boast. Nevertheless, most East Greenlandic men I met kept an exact account of the number of the animals they had caught and have a rough idea about the catch of other hunters. While hunters do not *openly* compete, people know about others' successes. This knowledge, in many cases, comes from observations and verbal exchanges. Very rarely does it come from listening to the self-proclamations of neighbours, and it is not set explicitly in contrast to other persons' less successful performances. Showing off the numbers of prey taken would also interfere with the mutual respect between the hunter and his prey, which one finds in East Greenland as among many other people around the circumpolar North. This respect includes the presumption that animals voluntarily give themselves up to the hunter, and that the hunter only kills the animals that willingly offer themselves.

Nuttall gives a similar account from a village in Northwest Greenland. He describes how knowledge of seal hunting comes from villagers' going down to the pier to welcome a returning hunter and to catch a glimpse of the number of seals caught. The news quickly spreads throughout the community and becomes part of collective memory. In a manner parallel to my experiences in East Greenland, Nuttall explains that

> reputations are established and publicly endorsed without, interestingly enough, harming the essential statement of egalitarianism – although privately, this may not be the case . . . Hunters do not brag about their success, which is not self-ascribed but expressed in an unspoken recognition by the community. This is not to deny that hunters vie with one another in a more implicit fashion, but to announce publicly that one is a 'big hunter' conflicts with the community norm.
>
> (1992: 140)

Likewise, stories about hunting trips disseminate information about people, their journeys, and their activities. On the one hand, they provide important knowledge for other hunters, and, on the other hand, they play a part in establishing the reputation of the people featuring in the story.

Stories circulate from teller to teller, and storytellers sometimes point out changes to the narrative structure and contents. Stories may be performed by anyone, yet tellers must follow particular rules, such as referring either to personal experience, to a person who has passed on the story to them, or to tellings in the past (cf. Bodenhorn 2004: 38; Morrow 1995). The storyteller takes over responsibility for what is recounted, at least for the way it is told, above and beyond its referential content (Brenneis 1987: 502; cf. Bauman 1977). The 'right' to the story is thus established in the performance (Ingold 1986: 228). It is important to stress that there is no essential story that is passed on from one teller to the other. Stories in themselves, as Ingold writes, are not 'passive objects of memory', and they 'are *not transmitted*' (2004b: 171, original emphasis). Rather, they always unfold in the relationship between storytellers and listeners. Storytellers only relay what they know their listeners will understand, and their performance, likewise, depends upon storytelling abilities, acts of remembering, and so forth. In this, Morrow's writings about her work with Alaskan Yupik storyteller Elsie Mather are of particular resonance. Morrow asserts that '[t]he most respected conveyors of Yup'ik knowledge are those who express things that listeners already know in artful or different ways, offering new expressions of the same' (1995: 32). Storytellers do not attempt to present a fixed interpretation to listeners. 'What the tale is about is implicitly relational; meaning is created in any listener's connections between a telling, a teller, a time, and a setting' (*ibid*: 40; cf. Mather and Morrow 1998). Stories thus change, depending on the audience and on the teller.[9] For this reason, I have paid particular attention to what stories in East Greenland may convey (e.g., shared values or expectations, or particular information), and to particular knowledge and guidelines shared between tellers and listeners, which often involve visitors and guests in one way or another. I will now turn my attention to the goods and material items that are given to visitors or guests, and that are brought along and handed out by them.

Material flows: food, presents, and money

Sharing in East Greenland involves a great variety of material items such as tools, equipment, clothing, foodstuff, presents, and money. When talking about sharing among the Inuit people, one of the most frequently treated themes is food. Across the circumpolar North, the sharing of country food,[10] and in particular meat, formed and forms an important part of daily social life, though the very elaborated patterns of distribution from the past are no longer practiced (e.g., Bodenhorn 1988, 1993, 2000b; Collings et al. 1998). These practices reinforce kinship and social relations, and at the same time they reflect an intimate relationship between humans and animals as well as the non-human environment (Bird-David 1990; Ingold 1986, 2000; Nuttall 1992).

Also in East Greenland, sharing *kalaalimernit* (Greenlandic country food), and above all the meat of seals as well as walrus and some whale species, still involves most households to a greater or lesser extent, though it is practised more irregularly than before and involves fewer people (Buijs 1993; Hovelsrud-Broda 1999a, 2009b, 2000b; Robbe 1975b). Products of fishing and gathering have never been redistributed to the same extent, but they were and still are shared out to visitors who are offered what is available (Robert-Lamblin 1986: 130). Today, sharing practices differ from household to household, and from time to time. Mirroring my experience, Buijs explains that '[m]eat is still taken to a father or a mother-in-law, but no longer after each catch, and the share is no longer the part of the animal traditionally apportioned to that person' (1993: 125). Shares (*miartsiat*) are nowadays often distributed among related families, and sometimes also non-relatives benefit from the catch – for example, disabled community members, widows, and other non-productive inhabitants (cf. Robbe 1975b: 219). Presents, and to some extent money, regularly circulate among related and non-related community members, both on a daily basis and on special occasions such as birthdays or Christmas.

Sharing, nevertheless, not only occurs between households, but also within the household as a unit (Hovelsrud-Broda 2000b: 197). Most households comprise of at least one hunter who brings the country food, one person who has a stable income from wages or transfer payments, and a woman who is able to process the catch.[11] With respect to sharing outside the household, Hovelsrud-Broda (2000b) distinguishes three different types: 1) inter-household sharing, 2) sharing between temporary hunting partners hunting polar bears, and 3) community-wide distribution. One form of inter-household sharing concerns food and presents offered to visitors and guests during invitations and feasts, on which I will focus later in this chapter. Money also has to be included here, though I will deal with it rather briefly and mostly with regard to its role in gambling. With respect to food, I mainly pay attention to the sharing of prepared foodstuffs which, according to

Bodenhorn (1988: 2), may not be regarded as 'shares' but rather as part of 'shared hospitality'. Shares, she argues, are defined as earned by participating in the procurement of food, whereas sharing denotes the process by which relationships are maintained. Similarly, the distribution of shares in East Greenland is denoted by the verb *miarpaa* (gives him/her something reserved for him/her), which is not used for offering food to visitors. The verb *miarpaa* and the associated noun *miartsiaq* are not exclusively used for shares of meat or country foods, and nowadays sometimes refer to purchases from the shop. Thus, *miartsiaq* was, for instance, used for groceries I contributed to the households I lived in or visited on a regular basis, and when initially I called these donations presents (*pisaarsiat*), I was corrected by household members.

Food as part of hospitality

As has been shown in the above examples, apart from being able to provide drinks, to offer a meal (usually of meat or fish, either Greenlandic or imported) forms part of the hosts' obligations and is the expected way to receive visitors and guests (cf. Robert-Lamblin 1986: 224). 'Whoever is hungry has to be fed', serves as an underlying premise, as my friend Sibora once explained to me. All visitors and guests may help themselves to tea or coffee,[12] and only sometimes are they specifically prompted to do so. As I have illustrated with the example of the disabled lady, I have repeatedly found that without asking, visitors would get themselves a cup from the kitchen cupboard and help themselves to tea or coffee. They also often enquire after it by asking '*Tsirapisi?*'/ '*Kappirapisi?*' (Do you [pl.] have any tea/coffee?). Additionally, bread is frequently volunteered, or likewise requested. These basic foodstuffs are available in most households most of the time, though they are imported and have to be bought in the shop (albeit relatively cheaply), and the right to their consumption on the part of visitors and guests is closely linked to other very basic aspects of hospitality that may be demanded (such as entering a living room or staying somewhere overnight). Being offered a proper meal, by contrast, is habitually preceded by an invitation from the hosts, and is not usually demanded by visitors.

To eat, visitors usually take a seat at the dining table, in the kitchen, or – less frequently – at the coffee table, where they consume their food without the immediate company of their hosts. Though sometimes visitors happen to eat together with their hosts, one does not encounter an expectation to be entertained or likewise to have to provide entertainment during meals. Robert-Lamblin writes: 'The value given to food is also expressed by the way visitors are honoured by being offered a large portion of meat, or the best one can find to offer them' (1986: 124). In the majority of cases, food offered to visitors consists of *kalaalimernit* (Greenlandic country food), especially in very rural places such as Sermiligaaq where most inhabitants have regular access to country food. *Kalaalimernit* holds much higher value than store-bought food and its consumption forms part of the identity of being Sermilingaarmeeq or Tunumeeq. Accordingly, inhabitants who did not know me well recurrently asked and tested my willingness and ability to eat and enjoy Greenlandic country food, which was then positively remarked upon. My own consumption of *kalaalimernit* was thus regarded as a marker of my adaptation to the Greenlandic way of living (on identity and Greenlandic food see Freeman 1996; Hovelsrud-Broda 1999b; Nuttall 1992: ch. 9). In Sermiligaaq, store-bought foodstuffs are generally less important for sharing, and on a daily basis they are shared with visitors only in households that – temporarily or more permanently – do not have access to country food.

The high value of *kalaalimernit* and its importance for sharing, is linked to the idea that the hunting territory and proceeds of a hunt are the common possession of all of the area's inhabitants (Buijs 1993: 121). Because animals give themselves up to the hunter, the hunter is obliged to distribute the meat to other community members. The respect shown towards animals includes the sharing of their meat, which again ensures a continued abundance of prey (Gessain 1969: 85). This idea forms part of accounts from oral history, such as the famous myth about the sea woman who keeps animals for herself due to humans' disrespectful behaviour towards animals and other human transgressions (e.g., Holm 1911; Gessain 1978; Sonne 1990). In summary, as Nuttall puts it:

> Free distribution is an acknowledgement of the debt owed to the animal in coming to the hunter and a denial that any one person has exclusive claims to ownership of the animals that are caught.
>
> (1992: 143)

Though a lot of the attitudes and rituals that acknowledge these presuppositions have disappeared in the course of time (see, for example, Kleivan and Sonne 1985),[13] practices of hosting visitors still clearly point to this understanding.

Store-bought foods (*qattunaamiit*), on the other hand, are treated differently than *kalaalimernit*. This becomes evident when guests stay for more than one or two meals. As has been illustrated in the story in which I travelled to Tiilerilaaq with my friend Hansigne, longer-term guests usually buy their own groceries and do not expect to be provided continuously with meals. I did not encounter any formal arrangements concerning the size of guests' contributions, but there are implicit expectations (cf. Bird-David 1987: 158). Hosts usually try to be as generous as possible, but in the longer term and in cases when, for example, they are themselves short of food, guests are sometimes expected to prepare their own food independently of the household members. Thus, during our stay in Tiilerilaq, our second elderly host had received only a small share of meat on the day of our arrival at her place, and was not able to offer us any of it. We consumed some of our store-bought food without including her. Hence, guests rarely share meals from store-bought foods with their hosts, though local meat or other *kalaalimernit* brought along by guests are usually consumed conjointly by household members, visitors, and guests. Groceries bought by guests are always treated as their property, and are neither used by hosts nor incorporated into their supplies. In Tiileri-laaq, for instance, Hansigne most naturally placed the supplies we had bought in the shop in one of our host's cupboards, and then removed all of them to take along when we left. Likewise, when I once stayed over at a friend's home in Tasiilaq and had bought some foodstuffs, I was explicitly asked to take along 'my' property, which I had meant to leave silently behind to contribute to my hosts' household.

One thus encounters different notions of property with respect to hunted and gathered foodstuffs and foodstuffs from the shops. The latter are to a greater extent treated as the property of a particular household or person, which in many cases does not imply the obligation to share with visitors or hosts. This distinction, nevertheless, is not absolute, since at times store-bought foods may also be incorporated into sharing practices and treated similarly to country food. They are sometimes shared out to visitors – for instance if a household does not have access to country food or if a lot is needed, such as when throwing a feast. One encounters differences along these lines between wage-earning and hunting families, and between town and village respectively.

It is important to add that sharing meals with visitors is not only expected by people visiting, but is also important for the hosts. For example, when I sometimes declined to join a meal at my host family's place since I had already eaten at a friend's house, my hosts were not very pleased. Likewise, other families I visited were often keen that I should eat with them. This can partly be explained, I believe, by the prestige generated by food sharing. The act is rewarding for the givers in that it demonstrates the hospitality and generosity of the particular family or person (cf. Ingold 1986: 228, on prestige-conferring generosity among hunter-gatherers). The sharing of substance when eating together, moreover, creates bonds between hosts and visitors or guests, and affirms and strengthens relationships. Likewise, offering food to visitors or guests provides a way to present wealth, in terms of the quantities of country food available (cf. Henriksen 1993: 51, on the Naskapi). In addition, sharing also implies a sense of community and some kind of altruism. This was noticeable, for instance, in the example of the disabled woman, who was pitied due to her illness.

How festivals promote sharing practices

Occasions that are especially important for sharing food and also presents include celebrations arranged for birthdays, confirmations, weddings, a child's first day at school, or a young boy's first catch. A present is called *pisaarsiaq* (a thing one has received), which stands in contrast to *miartsiaq*, which denotes a share given to somebody entitled to receive this particular share (e.g. piece of meat). Before the arrival of the missionaries, celebrations in East Greenland 'were occasional in the sense that they were promoted by particular events that were difficult to predict – such as the first steps of a child, the first seal or bear hunted by a young boy, the arrival of visitors, etc.' (Robert-Lamblin 1986: 134; on first catch celebrations see Nuttall 2000a). Holm reports from the end of the 19th century that any chance of feasting was taken, especially if it preceded long journeys, and the danger of being cut off (1911: 130).[14] These events, some of which have survived in slightly altered forms, have been complemented by various festivities from the Western calendar, both religious holidays and other important dates throughout the year.

Feasts organised for a particular person (or several persons) often take place at people's homes, though occasionally they may be celebrated formally in a reception room, especially in Tasiilaq. I have not experienced such formal celebrations on these occasions in Sermiligaaq. During my fieldwork, birthdays provided one of the most common reasons to organise a feast. Birthdays are not always much noticed, but if a family announces a celebration, then usually great numbers of villagers are invited, relatives and non-relatives alike. In West Greenland such celebrations are called *kaffimik*. Though this term is widely understood in the region, I have hardly ever heard inhabitants using it apart from Danish residents. In contrast to informal visits, feasts require an invitation.[15] Close relatives are usually told beforehand; at times they contribute *kalaalimernit*, cakes, and other foodstuffs. Invitations are usually distributed very shortly before the event (and sometimes even when it has already started), and often phone calls are made around the village. Such calls time and again proceeded quite similarly and in a brief manner: Without a word of greeting, a household member announces '*qaqqusiuu*' or '*qaaqqutsalu*' (we are celebrating/we are going to celebrate) and the time when the feast starts – in case it has not already started. The person answering the phone curtly asks for the name of the celebrated person – *kia?* (who?) – to be able to organise a suitable present. The person calling gives the name of the celebrated person, which is met with an 'OK'. Then the telephone is hung up. Not all occasions demand presents, but birthdays and confirmations do. Presents may include a variety of things, from implements, tools, toys, electronics, clothes, and cosmetics, to money. Events usually start at around lunchtime (sometimes earlier or later depending on the daily rhythm of the family members), and last until the evening. Within this time span visitors are quite flexible to choose a time that suits them, and there is continual coming and going.[16] Feasts may also be celebrated in the absence of the person concerned; for example, one day in 2007 a friend in Sermiligaaq had announced the celebration of her husband's birthday though he was away on a hunting trip.

During such celebrations, as with most community assemblies and festivities, the food, and in particular *kalaalimernit*, is highly important. Foodstuffs provided include both *kalaalimernit* as well as *qattunaamiit* (Danish products, store-bought food), yet most appreciated is a great variety of Greenlandic meat (*nereq*). Sometimes specialities such as *mattaq* (whale skin with blubber) or whale or walrus meat are presented. The food is often served on pieces of cardboard on the floor – in the kitchen or the entrance room – and visitors eat one by one without talking much (see Figure 5.2). The social part of the event takes place around the coffee table, where desserts, coffee, and tea are offered. Stories and jokes are told, and news and gossip is exchanged with great joviality. Most households make sure that they host their own feast once in a while, and that they have a great variety of *nereq* and other *kalaalimernit* available in order to improve their standing in the community. Well-off households are able to host feasts on a more regular basis, and are indeed expected to do so since they have the resources.

The procedures for such feasts are quite similar whatever the occasion, though there are minor differences with respect to the different events and households, the number and kinds of people invited, or the availability of presents. Birthdays

Figure 5.2 Birthday celebration, Sermiligaaq, April 2007
(Photo: Sophie Elixhauser)

and celebrations for a particular person often involve larger crowds of invitees from the community, related and non-related alike, whereas during primarily family-based festivities, as for example Christmas and Easter, mainly relatives are invited. I will give some examples.

Christmas

Christmas is preceded by a period of several weeks during which houses are prepared, foodstuffs are accumulated, and presents are organised. Most women spend a substantial amount of time cleaning their homes; pictures are removed, furniture is shifted and walls are wiped off and sometimes newly painted. In most living rooms the position of the furniture and the decoration of the walls consequently changes once a year. The night before Christmas Eve, homes are laboriously decorated with all kinds of Christmas decorations and colourful embellishments, including stars, religious pictures, and Christmas bowls. At my host family's place the night before Christmas Eve 2006, the parents and I worked from early evening until four o'clock in the morning in order to finish decorating. The children were not supposed to see us. The distribution and pooling of gifts is very important, and presents are often of high value. I was surprised by the amount of money spent on Christmas gifts, even by less affluent families. In the run-up to Christmas, the local stores are packed at almost every time of day, and huge parcels arrive via the local postal service with goods ordered by catalogue from Denmark and some other countries.

Early in the morning of 24th December, presents are exchanged among household members. Starting on 24th December, Christmas lasts for two weeks in East Greenland, as Sermilingaarmeeq repeatedly emphasised to me. Various village parties take place, and people make a lot of visits. Within these two weeks, related families distribute presents amongst each other; friends are usually not included in this practice. In order to deliver gifts, villagers walk to related families' homes, whereas when they themselves receive visitors they only accept but do not give. A visiting family group usually provides every household member with a present, including all of the children. It took me a while to understand these patterns. For example, for Christmas 2006 I had only received presents from families related to my host family, though I myself had bought presents for families and individuals that I often visited but that were not necessarily related to the family I lived with. Not yet aware of the local practices, in a number of cases I had given presents to people who did not give something to me, and likewise I had received gifts from others that I had not thought about.

Apart from presents, the availability of huge amounts of food is considered crucial during Christmas, both for household members' own consumption and in order to host visitors and guests. Apart from *kalaalimernit* and meals prepared from *qattunaamiit*, the masses of cakes, cookies, and sweets are particularly striking. Most households put up several bowls in the living room and kitchen filled with candy, gummy bears, or chocolates. Cakes are baked and provided in substantial quantities, and everybody may take as much as he or she likes. In order to further exemplify the importance of presents in East Greenland, I will elaborate upon the winter festival *milaartut* that takes place at Epiphany (*Kongepingasut* in Greenland).

Milaartut

The East Greenlandic *milaartut* starts on the evening of Twelfth Night, thereby concluding the two-week period of Christmas, and lasts the whole day of January 6th. It contains elements from both Scandinavian traditions and Greenlandic shamanic traditions (Nooter 1975; Nellemann 1960).[17] It is quite similar to *milaartut* customs in West Greenland, and reminiscent of related festivals among Inuit in Canada and Alaska (cf. Kielsen 1996; Kleivan 1960). The main event in the evening of January 5th is a dancing competition performed by a number of grotesquely disguised figures whose faces, blackened and distorted with plastic tape, resemble East Greenlandic masks. Before and after the competition at the *katersutarfik* (meetinghouse), or a comparable assembly hall, the *milaartit* (disguised players, sg. *milaarteq*) walk through the village and scare the inhabitants. They mostly appear in pairs, and they are not allowed to speak.

The disguised players usually visit different houses in the settlement. In Sermiligaaq in 2007, on the evening of January 5th, a number of adult *milaartit* – mostly young and unmarried people – were seen outside in the village. Through their outer appearance, strange movements and attempts to scare the inhabitants, they created an atmosphere of fear, combined with an element of 'frighten and amuse', as Nooter (1975: 165) calls it. During Epiphany itself, the children dressed up and walked from house to house, sometimes in the company of a parent or other adults; I accompanied my best friend and her dressed-up children. We visited several houses, which mostly but not exclusively belonged to related families. The household members appeared frightened when seeing the *milaartit* (though quite apparently in a half-serious way), and addressed them as strangers. Nonetheless, the disguised children and the accompanying persons were asked to come in and were sometimes offered a drink or some snacks. The main activity of the inhabitants of the houses was to identify the dressed-up children and, after that, to give them each a present. Presents included a wide range of things, from a handful of sweets, a can of peaches, or other foodstuffs to more valuable goods, such as a CD player that was given in one instance. Some families thereby disposed of superfluous household items. For instance, my friend Maline's three-year-old daughter was given a cooking pot from the kitchen and some decorative clay figures.

During my time in Sermiligaaq I have only seen children walking from house to house and receiving presents (see Figures 5.3 and 5.4), yet Tunumeeq have explained to me that both grown-ups and children may practice this custom (cf. Nooter 1975). The dressed-up players may also demand gifts, and these requests may not be declined. In this way, from time to time very expensive items are given away, such as a winter coat – which an elderly lady had originally received in this same way, as she told to me – and sometimes even babies. For these reasons some residents are not willing to participate in this custom, and, as a woman from Sermiligaaq told me, each household has the right to put a cross on the wall of their house, which signifies that the residents do not want to join. I have not, however, observed this kind of demand sharing or such crosses for myself, nor have I come across cases of the gift of babies during *milaartut* (Nooter 1975: 160, mentions one such case from Tiilerilaaq in the 1960s). Such cases seem to be very rare nowadays, and I am not sure if the custom of demanding babies is still practised.

As Nuttall has observed, with respect to similar customs in Kangersuatsiaq, West Greenland, the *milaartit,* on the one hand, symbolise strangers from the community who scare the inhabitants, but on the other hand their visit 'also emphasises core moral, cultural and religious values of hospitality' (1992: 115), in terms of particular obligations. 'In treating the stranger as a guest, . . . fear is dealt with because the ritual and moral nature of hospitality acts to establish temporary bonds commensurate with kinship, friendship and obligation' (*ibid*). Thus, I suggest that the importance of hosting and distributing presents to the disguised players lies in the underlying premise that sharing, as an obligation and

Figure 5.3 Child *milaarteq*, Sermiligaaq, January 6, 2007

(Photo: Sophie Elixhauser)

Figure 5.4 Child *milaarteq*, Sermiligaaq, January 6, 2007

(Photo: Sophie Elixhauser)

an important aspect of hospitality, also applies to strangers. Holm has mentioned the same point:

> The mutual hospitality of the natives knows no bounds. It is not counted as a virtue by them, but as a stern duty. They are hospitable to all without exception, and always share with one another, even with strangers, till they have reached the end of their supplies.
>
> (1911: 137)

With some qualifications, this statement still applies today. East Greenlanders often show great hospitality towards strangers without any expectation of return (cf. Thalbitzer 1941: 641). Nevertheless, due to the flux of strangers, either in residence (mainly Danes) or on holidays (tourists of various origins), and the experience of their different habits, things are changing, and many East Greenlanders are beginning to express reservations.

Let me return to gift giving during *milaartut*. Information about the amount, value, and kinds of presents a person or family has received or given during *milaartut* circulates among community members as part of everyday conversations. As with food and presents distributed on other occasions, giving out presents during *milaartut* involves some element of 'prestige-conferring generosity' (Ingold 1986: 228). The more a person has to offer, the better; and the more generous they appear to be, the higher the prestige. The prestige involved, nevertheless, is limited due to the underlying principle that recipients are entitled to receive the things they are given (Peterson 1993). Givers have to hand out gifts on demand to *milaartit*, and in this they have no choice. Yet they do have some choice in the size and value of the gift, which again might be remarked upon positively (or negatively) by others.

The value of presents

In relation to this, I want to dwell upon my impression that in many situations the act of giving as well as the size and the value of a particular item seems to be considered more important than the item given as such and its usefulness. Though of course East Greenlanders value particular things over others, attitudes towards items bought from the shops are often rather casual, and no one appeared too disappointed if, for instance, a toy broke down, or if other goods turned out to be of no great use. Likewise and as mentioned above, attachment to residential buildings is not very strong, and moving house did not usually involve taking along a great number of personal items. By contrast, presents manufactured by the giver herself, as for example handicrafts such as knitting or pearl embroidery, seem to hold a higher value, a difference which parallels that between the valuing of *kalaalimernit* and *qattunaamiit*, as mentioned above. The difference lies in the bond between the producer and the produced, which is established through the production process in the case of handicrafts, or in the particular relationship between the hunter and his prey in the case of *kalaalimernit* (cf. Ingold 1986: ch. 9; Riches 2004). Thalbitzer writes that 'the spiritual connection with the object owned' is a crucial point in this respect and that it 'plays a still greater role as regards things which the owner himself has made, or has acquired at great cost. A person tears off a part of himself by ridding himself of his accustomed property' (1941: 640). This argument still seems relevant to the explanation of some contemporary attitudes.

On a broader level, the relatively small importance of many personal possessions, which was a precondition for the mobile lifestyle of former times, is to some extent still observable today. In addition, many Tunumeeq I met are particularly interested in, and much aware of, the price spent for a present by the giver, and sometimes also directly ask about it. This is not considered impolite. There are particular expectations concerning a present's size, though if expectations were not fulfilled a person would rarely say anything or complain about it. Just as Morrow has remarked with respect to the Messenger Feast, a festivity among Alaskan Yupiit, 'The burden was on the giver to determine, based upon previous gift-exchanges, years of observation, and specific knowledge of people in the community, what constituted an appropriate gift' (1990: 149). At the same time, presents in East Greenland may also be demanded, similar to how food is occasionally requested. These requests, however, are usually framed rather carefully in terms of a question or suggestion in order not to encroach upon the other person's autonomy, just like the invitations described above. I have also experienced demand sharing directed towards myself, and I was sometimes prompted to buy goods from the supermarket or to bring particular presents upon my return from Germany. I usually tried to follow such suggestions, yet when I was sometimes unable to do so my friends would usually accept my explanation.[18] I was also often given presents myself. This only sometimes took place in relation to the presents I had brought along, and I did not have the feeling that reciprocity was expected. These observations parallel Bird-David's and Ingold's observations about hunter-gatherers who see their environment as 'giving'. 'Practically, would-be-recipients request what they see in the possession of others and do not request them to produce what they do not appear to have' (Bird-David 1992: 30, cited in Ingold 2000: 71). Accordingly, Tunumeeq usually ask for things that others have or that they assume others might have or are able to obtain. What a person is willing to give is usually accepted, though in the long run, a person's failure to share out things she most obviously appears to have may lead to grumbling.

Such behaviour is characteristic of what Woodburn (1982) calls an 'immediate return' system among hunter-gatherers. The category of 'immediate return' is applied to egalitarian hunter and gatherers, and is set in contrast to 'delayed return' economies among non-egalitarian groups. Societies in the first group practice little or no storage, and consume food and other resources immediately after their procurement. They regard the environment as a full storehouse, which will always provide. Because of this, and contrary to delayed return systems, Woodburn argues, there is a lack of investment towards particular persons and resources. Delayed return hunters and gatherers, on the other hand, store surplus food and tend to accumulate property. My East Greenland research shows some characteristics of the immediate return economy, as for example in the obligation to share out foodstuffs and presents along with the attachment of a relatively low value to the accumulation of personal possessions. Such sharing practices, which are not necessarily based on reciprocity, stress mutuality and do not tolerate inequality. Nevertheless, by Woodburn's argument, East Greenland should also be unequivocally delayed return. People invest in substantial items of equipment, such as boats, and they practice extensive storage. Hence, Woodburn's dichotomy is rather difficult to apply in this context.

Mutuality and a lack of tolerance for inequality are apparent not only in the practices of sharing food and presents, but also in the distribution of money during gambling. In the following, I will treat gambling, which in Sermiligaaq and other villages in the region most often takes place in people's homes, as another form of visiting, which involves people entering and leaving a house as well as sharing particular goods, in this case money.

Gambling and the redistribution of money

Practices of pooling and redistributing stakes during gambling are deeply embedded in inhabitants' broader attitudes towards money. In contrast to foodstuffs, and especially *kalaalimernit*, which in many contexts involve the obligation to share,

I did not come across generally accepted rules concerning the sharing of money (*aningaasat*). Robert-Lamblin explains this in a similar manner:

> No rules have been established for distributing money earnings among relatives in the newly-developed occupations like commercial fishing, handicrafts or wage-earning activities. Although one receives some portions of meat as gifts though the traditional system of meat sharing, it is completely accepted that one keeps money one has earned for oneself.
>
> (1986: 131)

Though money donations or transfers are quite common among households and related families, all in all, family solidarity demands less of people who live on monetary resources alone than it does, for instance, of hunters (cf. Riches 2004, who reports similarly with respect to Canadian Inuit as well as hunter-gatherer groups elsewhere). Robert-Lamblin regards this as one of the reasons for the establishment of a middle-class bourgeoisie in East Greenland, whose members have accumulated the financial means in order to take advantage of material comforts, trips abroad, and so forth (1986: 131). Nevertheless, money is not totally excluded from sharing obligations, which becomes clear when looking at practices of gambling.

Gambling is a very common spare time activity among Tunumeeq in general, and Sermilingaarmeeq in particular. It is practiced during both winter and summer, though it is more widespread during the winter months. The proportion of women is higher than that of men, and men usually spend more of their leisure time outside. Gambling has a rather bad reputation, and though various villagers participate in it once in a while, it is often frowned upon. Sermilingaarmeeq often discuss how often which person has won or lost, and are very well informed about these daily details. The most widespread game is bingo, played either with numbered tiles or with playing cards; additionally, several other card games are practised. Habitually, different games are played in the houses of different villagers, and the locations regularly change amongst the villagers' houses. During one week X's house is the meeting point (and additionally place Y for those wanting to play another game than that practised at place X), whereas the next week everybody gathers at place Z. Bingo events also sometimes take place at community buildings, though in Sermiligaaq this only exceptionally happened during the time of my fieldwork. In Tasiilaq this is much more common. These parties habitually start in the early afternoon and last until the late evening. News about the gambling locations quickly spread among the villagers. Tea and sometimes coffee are offered but usually no food. At the outset, hosts usually organise the minimum number of people needed for a specific game, which is often quickly added to by other villagers dropping in.[19] There is a flux of people entering, leaving, and re-entering, some just in order to take a look or to leave a message for someone, whereas others stay for many hours. The living rooms of such houses are often packed with people and gamblers ranging from young adults to elderly inhabitants occupy every little piece of table, seat, and floor. A female villager once jokingly stated '*tattanni klubi anngiitseq*' (this is the club for the adults). These meetings provide an occasion to exchange the latest gossip and to get to hear the village news. Lots of joking takes place and the villagers enjoy the communicative circles and the social atmosphere.

The husbands of the bingo-hosting households, especially those who do not gamble by themselves, are often less delighted by the crowds of people occupying their homes. During gambling sessions, many of them tended to prefer staying outside or visiting somewhere else. After a few days the location usually changes; hosts have become tired of the crowds of people around, and household members have started to claim back a peaceful living room.

I, too, sometimes joined these parties in Sermiligaaq, though usually after one or two hours my excitement had somewhat lessened and I tended to stop playing. Such behaviour was quite the exception, since most people stayed for many hours, and only went home early if they ran out of cash or needed to see to other important obligations. Some people would borrow money from other participants to be able to continue playing. On occasion, housewives would be caught up with this activity that they forgot about duties such as cooking or looking after the children. It was not unusual for a husband at home alone, babysitting, to call the 'bingo house' in search of his wife.

Most probably, the main reason why gambling is often frowned upon is the financial aspect. Gambling often involves substantial amounts of money, and repeated losses can result in financial problems. Stakes differ from location to location as well as in the different settlements in the district. For example, when joining the bingo circles during my few days' stay with Hansigne in Tiilerilaaq, stakes were double compared to what we were used to in Sermiligaaq. In Tasiilaq I have heard of even higher amounts.[20] Robert-Lamblin also reports about items and goods that were staked, yet among men only, something which I have not encountered myself. She also writes that stakes were higher among men as compared to the women, which I have not noticed (Robert-Lamblin 1986: 134). Quite to the contrary, in Sermiligaaq the most prominent players were female. In general, I got the impression that, as the game is based on chance; those people who regularly gamble seem to win as often as they lose. East Greenlanders do not seem to join this activity for the money they are hoping to win. This has also been observed by Robert-Lamblin:

> Since the gambling partners are usually the same people, what is lost one evening can be won back the next, and for the most part the money and goods circulate within the same circle of gamblers.
>
> (*ibid*)

Though financial gains are appreciated and are sometimes positively remarked upon by other household members, an important incentive for most people seems to be the communicative momentum. Riches argues similarly with respect to the Port Burwell Inuit in Canada:

> There are many motives for joining a party. The desire to win sorely needed cash may be one, yet I have been to lively sessions where bullets only were in circulation. It is perhaps better to view participation primarily as a

manifestation of community integration: through joining a game a resident may assert membership, and a stranger, incorporation, in a community that otherwise lacks mandatory work or ritual group.

(1975: 26)

Hence, one of the broader functions of gambling events in Sermiligaaq is that they strengthen ties through mutual commitment, and bring about community cohesion (cf. Woodburn 1982: 444). Family members, for instance, often form teams and give out shares of their wins to other persons of their family group, and sometimes also to friends who have run out of money. Likewise, during one initial party I joined, in which I had repeatedly been losing, some participants were worried that I might not enjoy myself, and tried to support me by sitting next to me in order to keep an eye on my tiles and sheet of paper. Through helping me to win a game they were hoping that I would stay longer and would return the next time and entertain everyone. Another observation that may be understood in terms of community integration is the communal laughter and amusement induced by using nicknames for specific playing cards, denoting village residents well known by other villagers. Many residents receive a nickname (*taagulaq*) in addition to the names bestowed at baptism (cf. Robbe 1981: 53). Such a name may stay with a person for a long or short period of time; it may sometimes fall out of use after a while or be replaced by another one. Nicknames used in a particular settlement are not necessarily known to the inhabitants of other settlements.

A common rule I had to learn whilst participating in these games is that after winning a substantial amount of money, participants are not supposed to leave the round and take the money home. Other people present expect this person to continue playing, and to restake the accumulated money in order that it should be redistributed again among the participants. The first few times I joined the card games and managed to win, without greatly reflecting on it I usually thought to leave soon afterwards. My fellow participants, however, did not approve. They would somewhat critically ask me why I would want to leave at that particular juncture. In these cases the accumulation of individual gain is not appreciated. Social pressure to continue playing and to stake what is won has the effect that other participants are able to regroup their lost possessions. Gambling thus works towards perpetuating equality in the distribution of monetary returns, and operates against the possibility of systematic accumulation (Woodburn 1982: 443; cf. Mitchell 1988, for an example from Papua New Guinea). The East Greenland case supports Per Binde's statement:

Egalitarianism is expressed by morals, norms, and customs that regulate the behavior of individuals; sharing is typically encouraged while egoism is condemned, and attempts at coercive domination are criticized or ridiculed.

(2005: 453)

Such characteristics are also reflected in many East Greenlanders' attitudes towards money more generally. Though notions of saving or postponing purchases may be encountered with respect to those inhabitants who have attained a certain level of income, the habit of spending money immediately on receipt is still widespread.[21] This becomes most obvious when looking at inhabitants' purchasing capacity on pay days, on which most shops are packed and great sums of money pass the counter. One way of dealing with situations when there is no money left is to try to borrow from those who save up, or receive payments on different dates. Borrowing is very widespread in Sermiligaaq and other places in the region.

Conclusion: sharing, property, and trust

Sharing defines to whom you are related, and is as an expression of community. These relations, which may be considered in terms of property or entitlements, regulate the use and the disposition of goods, and involve responsibilities as well as claims. For example, stories may be shared only if rules are followed on properly referencing their relational context. Sharing Greenlandic meat involves the responsibility to give away freely what one has received freely, whereas money is subject to similar obligations only to some extent and in particular situations, such as during gambling. Houses are claimed to be inhabited by particular household members, yet these claims do not deny others the right to enter freely and to use this living space. People's habitation of a house thus entails the responsibility to host other people who decide they want to visit. Hence, differing notions of property in contemporary East Greenland influence how particular goods are dealt with in terms of sharing. In this, changes from a predominant focus on collective property rights to an increase in more exclusive personal property rights with respect to various (mostly) imported items have to be taken into account. Accepted systems for redistribution still persist, above all with respect to long existent resources, but only to a limited extent do they take into account goods introduced rather recently. Along these lines there are differences – for example, in the treatment of *kalaalimernit* as opposed to *qattunaamiit*. The former involves sharing obligations with respect to visitors, and likewise hosts, whereas the latter often does not. Such distinctions, however, provide vague guidance only, since on the one hand imported items are sometimes incorporated into sharing practices (e.g., the use of houses, gift giving), and on the other hand obligations regarding the distribution of Greenlandic resources (e.g., meat) have decreased over time and are no longer practised to the same extent as before.

Several examples in this chapter draw attention to characteristics that have been observed among other hunter-gatherer societies with an immediate return economy. Though the last century has brought great changes to the hunting and (to a lesser extent) gathering economy and society of East Greenland, features such as the low value attached to personal possessions, poorly developed saving practices, and the importance of sharing, especially with respect to visitors and guests, are still prominent, at least among parts of the population. The latter strongly point to what Sahlins has called 'generalised reciprocity'. East Greenlanders are obliged to host more or less any person who asks for it, and though this is sometimes reciprocated in kind, it is done without calculation of return or expectation within a particular period of time. In fact, visitors regard hospitality not as a voluntary action on behalf of their hosts but as an entitlement on their own behalf. Yet this entitlement also implies particular expectations for visitors and guests, such as providing entertainment (e.g., sharing stories), and contributing food if staying for a longer period of time. Characteristics of an immediate return economy are especially evident with regard to the emphasis on personal autonomy, in all fields of activity and relationships. These characteristics, nevertheless, are accompanied by others which conform to Woodburn's description of a delayed return economy, as I have explained above. So how do we account for the presence of these immediate return features, given that the East Greenlandic economy, in Woodburn's terms, is unequivocally of the delayed return type?

I believe that the resolution of the problem lies in Woodburn's definition of personal autonomy. Woodburn has argued that autonomy has to be seen as a limiting factor for investment in and dependency on others. According to him, immediate return economies are characterised by a lack of 'long-term binding commitments and dependencies' towards particular resources or people, in contrast to delayed return economies characterised by such dependencies (Woodburn 1982: 434). My East Greenland research, however, shows a strong link between autonomy and long-term commitments to particular people and resources, and underlines Ingold's argument that lack of commitments is not a necessary condition for personal autonomy. Ingold argues that sharing practices among hunters and gatherers may better be viewed in terms of *trust*. He explains:

> The essence of trust is a peculiar combination of *autonomy* and *dependency*. To trust someone is to act with that person in mind, in the hope and expectation that she will do likewise – responding in ways favourable to you – so long as you do nothing to curb her autonomy to act otherwise. Although you depend on a favourable response, that response comes entirely on the initiative and volition of the other party. Any attempt to impose a response, to lay down conditions or obligations that the other is bound to follow, would represent a betrayal of trust and a negation of the relationship.
>
> (2000: 69–70)

This combination of personal autonomy and dependency on others is evident in most of the examples presented. East Greenland practices of visiting and being visited entail a reliance on others, yet this reliance goes hand in hand with great carefulness not to impose on them particular wishes and expectations. This is highlighted through various communicational modalities. Furthermore, though people's entitlements in terms of shared hospitality are sometimes demanded, such demands are often framed in a way that leaves the final decision up to those of whom they are made, which ultimately protects their autonomy.

This notion of trust, which should not be confused with confidence,[22] applies not only to relations among humans, but may be extended to include relations with animals hunted or with other non-human actors in the environment as well. Trust, moreover, always includes an element of risk that 'the other on whose actions I depend, but which I cannot in any way control, may act contrary to my expectations' (*ibid*: 70; cf. Gambetta 1988; Corsin Jiménez 2007). This may result in suspicion towards fellow inhabitants, which I also repeatedly experienced in East Greenland. Amongst other things, Sermilingaarmeeq pay close attention to what gifts are given to whom, and one still sometimes encounters suspicion towards those who most obviously have resources or other goods at their disposal but who do not share. This has also been described by Campbell Hughes, who mentions Rasmussen's observation concerning 'the suspicion of other people's behavior behind the overt hospitality shown by hosts' (1958: 373). My research, then, supports Ingold's argument that lack of commitments to, or investments in, specific others is not, as Woodburn believes, a necessary condition for personal autonomy. On the contrary, autonomy can be sustained despite such commitment and investments, precisely because relationships – with both humans and non-human entities – are based on trust.

In sum, the giving of gifts or food, and other types of sharing, not only reduce inequality – as Robert-Lamblin asserts, referring to times in East Greenland before the major economic and cultural changes brought by colonising powers had taken place – but 'may also have been due to the wish to avoid envy, jealousy or dangerous rivalry' (1986: 145). She continues:

> If not too intense, they could be solved by humour, laughter or jokes. But when tensions ran high (as for example abduction of a wife by another man) the "duels of song" was an effective way of settling conflicts while usually avoiding bloodshed.
>
> (*ibid*: 145)

But what is the contemporary situation in this respect, and what has changed? What types of community sanction and means of social control are to be found today? This is the theme to which we turn in the following chapter.

Notes

1 This fluidity of group composition and constant movement of people has been recognised as one of the most characteristic features of hunter-gatherer societies (Lee and DeVore 1968: 7).
2 One encounters differences in inhabitants' statements about future plans with respect to deadlines and appointments that relate to the institutional context of the social system, such as regarding opening hours of shops, times when children must go to school, helicopter schedules, or the like, as opposed to plans that are not dependent upon any of these prescriptions. The former have been introduced only some decades ago.
3 Other common expressions include *tagitsau* (we will see each other), *tagivakkit* (I see you, sg.), *tagitsavakkit* (I will see you, sg.), and *ajinngitsaulit / ajinngitsausi* (take care, sg./pl.).
4 For instance, when I once passed by a friend's house in order to visit, and this friend was busy sweeping the floor, she told me to come back another time.
5 The municipality provides home helpers that support elderly community members in their daily tasks such as cleaning the house, filling up the water tank, and so forth.

6 Robert-Lamblin likewise remarked in the 1980s that 'the nomadic instinct has not totally disappeared among the other more sedentary Ammassalimiut' (1986: 50). Mentioning other 'less sedentary' inhabitants, she refers to a few hunting families who participated in a government-led initiative in the late 1960s to mid 1970s aimed at supporting long-distance migration. These families were brought to distant hunting areas for a one-year period to be able to perpetuate hunting life. Apart from support in transportation, they were granted loans to buy the goods indispensable for their stay. East Greenlanders nowadays talk very positively about these initiatives.

7 In her book about the life histories of three Inuit women of different generations, Nancy Wachowich (2001: 3) shows that in the Canadian Arctic stories remain important even today, though the themes and shapes of oral traditions have partly changed. This also applies to East Greenland, where one encounters differences in storytelling among people from different generations (a point I cannot further deal with in this book, however).

8 Cf. Basso's work on the Western Apache, and especially his article 'Stalking with stories' (1984), in which he shows that stories serve as a reminder of community expectations, and of the consequences of deviation.

9 Ingold ascribes some kind of subject character to stories. 'As manifestations of persons, stories are not objects of memory but living subjects with whom listeners can engage in acts of remembering'. He further argues that 'there is a certain sense in which stories are not *about* persons. Rather they are persons, who, in the telling, are not just commemorated but actually *made present*, through the mimetic reenactment of their voices and deeds, to the assembled audience' (Ingold 2004b: 171).

10 The term country food subsumes all foodstuffs which stem from the Greenlandic environment, and which are usually procured through subsistence activities, such as hunting, fishing, and collecting berries or plants.

11 In the Ammassalik region, the gendered division of productive activities is still very strong, and only exceptionally have I encountered men who processed their catch (cf. Hovelsrud-Broda 2000b: 195). This is quite contrary to the situation in Ittoqqortoormiit and many places in West Greenland, where nowadays the hunters themselves cut up the prey and process skins, amongst other things.

12 Coffee and tea, just like most other imported items, have been introduced rather recently to East Greenland. Robert-Lamblin writes that coffee was forbidden at the 'beginning of civilization' (i.e. after East Greenland has been missionised in 1894), and that this was still the case in 1940 (1986: 123–4). Today, the consumption of both tea and coffee is high, and black tea with sugar is consumed on a daily basis in almost every household.

13 One often reported hunting ritual included pouring fresh water into the mouth of a caught seal before cutting it up. This ritual act showed gratitude towards the seal for giving itself up to the hunter (Petersen and Hauser 2006: 18–9).

14 Celebrations at the time also sometimes involved the famous 'lamp extinguishing game', which included the exchange of sexual partners (Holm 1911: 69; cf. Thalbitzer 1941: 667–8).

15 The story depicting a visit during Vittus's birthday that I have recounted in the first part of this chapter does not represent such a feast. In this case, apart from the visitor from Kulusuk, only close relatives visited and the available food was limited.

16 A West Greenlandic friend from Nuuk once mentioned to me that Danes and foreigners sometimes misunderstand this practice and stay from beginning to end. However, they are expected to leave after having finished eating and taken some dessert and coffee, in order to make space for newly arriving visitors.

17 In this chapter I will consider *milaartut* mainly with regard to the distribution of gifts. For more information on the ritual, and especially the role of humour, fear, and the audience, see Chapter Seven.

18 For example, on one occasion I had temporarily run out of cash, and was not able to buy groceries (yet I was able to make up for it at a later point in time). Another time I was asked if I would want to give away a jacket I was wearing, which however I still needed for my own use.

19 At times children are also taken along, though most people try to show up without. These instances were among the few occasions in which I have heard Sermilingaar-meeq complaining about the presence of children.

20 The differing values of stakes in the settlements in the district also provide occasion for gossip among the inhabitants. For example, I have heard (half-serious) complains that Tasiilarmeeq play Bingo much too fast, that they bet too much money, and so forth.

21 Though less frequently, one still finds cases that fit the following description from Robert-Lamblin: 'Most East Greenlanders . . . live from day to day. They spend with pleasure and do not count the money coming in; then they look for an emergency solution to difficulties (selling a hastily-made handicraft item, attempting to catch fish, borrowing money, etc.)' (1986: 129).

22 Trust, as Ingold explains, should not be mistaken for confidence. The problem with confidence is that 'it presupposes no engagement, no active involvement on our part, with the potential sources of danger in the world, so that when trouble does strike it is attributed to forces external to the field of our own relationships' (2000: 71). By contrast, he continues to explain, 'Trust presupposes an active, prior engagement with the agencies and entities of the environment on which we depend; it is an inherent quality of our relationships towards them' (*ibid*).

6   Social sanctions

The balancing of personal autonomy and community expectations

June 21, 2007. Just like every year, some hundreds of East Greenlanders from all over the region have come together to celebrate the summer solstice at Ittimiini, a small peninsula at the outskirts of Tasiilaq. Some *qattunat* (Danish persons, sometimes used for Whites or Southerners more broadly) are also part of the crowd. People have spread out on the grass, eating, chatting, and enjoying the social atmosphere. Family groups brought picnics consisting of *kalaalimernit* (Greenlandic country food) and other specialities. A visiting British friend of mine and I joined my host family from Sermiligaaq and their Tasiilaq relatives. There is continual coming and going, and we meet several other Sermilingaarmeeq. Colourful nylon tents are scattered around the area, and near the outer edge of the peninsula two traditional summer tents made from wood and seal skins have been put up. In the afternoon a group of women, some of whom are wearing traditional dress, take the large *umiaq* (women's boat) from the Tasiilaq museum for a tour around the fjord. At some point, a female community member gives a speech in East Green-landic, Danish, and English about East Greenlandic cultural traditions. The Tasiilaq fire brigade show their skills; local rock bands perform; and some young people play traditional games, such as tug of war and football with a stuffed seal fur (see Figure 6.1). A dance at the sports hall in town has been announced for the evening activities. On arrival, I am told that one event I just missed was traditional song and drum duels (*iverneq*) performed by an elderly lady and some girls. I have seen similar performances on several other occasions during larger community celebrations or tourist events. Only few Tunumeeq still know how to perform a song and drum duel, and some of the schools have started to offer courses to keep the tradition alive.[1]

The duels provide an outstanding example of an institutionalised way of dealing with conflicts in pre-Christian times, which illustrates how East Greenlanders would communicate particular values and expectations. In this chapter, song and drum duels serve as a historical metaphor to illustrate features and ways of communicat-ing, which still characterise many informal communicative exchanges in people's daily lives. By and large, contemporary institutionalised mechanisms of conflict resolution follow the Western model, based on different premises than the song and drum duels and many East Greenlandic forms of interpersonal communication.

Figure 6.1 Young people playing tug of war, Ittimiini, Tasiilaq, June 21, 2007
(Photo: Sophie Elixhauser)

Rather than looking at these mechanisms and institutionalised settings such as schools, offices, local politics, and so forth, I focus here on communicative modalities used informally, which allow people to show that something is not as it ought to be. I will draw attention to the fact that people quietly, subtly, and indirectly control each other's behaviour in day-to-day life by means of specific ways of talking about self and others, ridicule and teasing, gossip, and withdrawal. A closer look at these so-called informal social sanctions reveals that some main characteristics of the institutionalised song and drum duels from before are still relevant in contemporary East Greenlandic society. Building on examples from previous chapters of the ways in which people are taught about values and expectations, such as teasing as part of the education of children (and adults) and practices of storytelling, the following chapter will broaden my analysis of these processes by focussing on examples from social get-togethers and gatherings at village meeting places.

Song and drum duels: dealing with conflicts in the olden times

In pre-Christian times, drum and song duels (*ivernit*, sg. *iverneq*) provided an important means and the only formal procedure for dealing with conflicts (Holm 1911; Gessain and Victor 1973; Rosing 1969; Smidt and Smidt 1975; Victor et al. 1991).[2] Motivations to hold them include envy, theft, and, as Holm writes, revenge 'for some wrong a man has suffered on a woman's account, as, for instance, when someone has run away with his wife or been too intimate with her' (1911: 128). More serious crimes, such as murder, were also sometimes dealt with in this way,

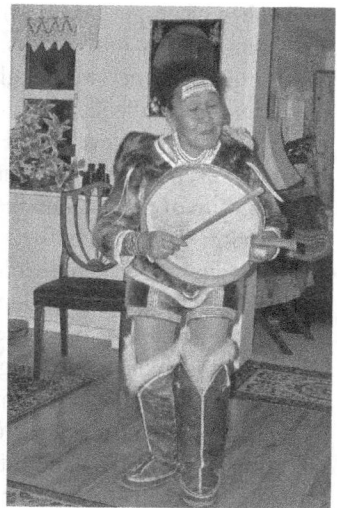

Figure 6.2 Drum and song performance by Anna Kuitse Thastum, Kulusuk, September 2008
(Photo: Sophie Elixhauser)

Figure 6.3 Drum and song performance by Anna Kuitse Thastum, Kulusuk, September 2008
(Photo: Sophie Elixhauser)

though on occasion they would result in blood feuds (Sandgreen 1987; Sonne 1982). Today, song and drum duel performances are occasionally carried out for purposes of entertainment and cultural revitalisation, but they no longer fulfil judicial purposes (see Figures 6.2 and 6.3).

Song and drum duels always involve two antagonists of the same gender. They are performed by both women and men, and, in former times, the audience consisted of supporters of both adversaries. The big gatherings in the summer provided popular occasions for these duels, as, for example, the large meetings at Qinngeq, a fjord near Kuummiut, where the inhabitants went each spring in order to catch *ammassat* (small capelins) and to socialise after the end of the long winter. Qinngeq has nowadays lost its importance, though a few Tunumeeq, especially hunters and fishermen, still go there each spring in order to catch *ammassat*. As Robert-Lamblin writes, 'Tasiilaq of today has replaced the Qinngeq of the past' (1986: 84). After the missionaries established Tasiilaq as the centre of the region, song and drum duels were also performed at Ittimiini, which is being re-enacted today (Gessain and Victor 1973: 147).

In a duel two offended parties exchange scathing songs (*pisit*) in the presence of an amused audience. The *pisit* are based on irony, sarcasm, metaphors, and insinuations, and they are accompanied by the rhythm of the drum (*qilaat*), played by the adversaries. The songs do not unambiguously focus on a conflict, but conceal it with ridicule and ambiguity, and often they carry sexual reference (for a collections of songs, see Victor et al. 1991).[3] The opponents stand close to each other when performing their songs, which are supported by gestures, grimaces, and bodily movements. Whilst an individual is singing, the other person stands still, or sometimes laughs, in order to show his or her indifference to the singer and the audience. Once in a while, he or she receives blows and cuffs to the cheekbones and forehead (Holm 1911: 128; Petersen and Hauser 2006: 68). In the old times, punishment occurred through the humiliation of a person in front of a large group of people, and the songs were sometimes rehearsed several months ahead. In this manner, conflicts were openly addressed, though the rules of a drum and song duel prevented a violent clash between the singing parties. The singing parties did not engage in arguments, and they could not provide self-justification. Indeed, the songs of the two adversaries often focused on different themes. It was up to the audience to judge a conflict through their laughter, shrieking, and attention, and questions of right and wrong often took a backseat. Judgement was based on the opponents' artistic performances. 'Drawing the community into the accusations serve[d] . . . both to give strengths to the accusations and to remove part of the responsibility from the individual accuser', as Penelope Eckert and Russel Newmark have argued with regard to song and drum duels in the Central Arctic (1980: 200). Occasionally, the duels were continued at a later point in time, and sometimes they were repeated merely for pleasure. Petersen and Hauser write:

> It is said that after the strife the opponents often became close friends for the rest of their lives and repeated the drum-song and dance every year to confirm their friendship and the joy of having solved a problem.
>
> (2006: 41)

Hence, apart from their punitive function, the duels could also strengthen friendships between the opponents (and between people from different settlements).

Inge Kleivan reports (albeit with reference to West Greenland) that satirical songs were also used in the battle against the Europeans, arguing that, 'The songs were to a certain extent probably responsible for the fact that the colonisation of Greenland was effected as peaceably as it was, because disagreement and aggression found expression in sharp words in this institutionalised manner' (1971: 16). Song duels thus provided a means both to bring interpersonal and intergroup antagonisms out in the open and to prevent more overt forms of hostility. The material from East Greenland illustrates that, just as Briggs writes with respect to Canadian Inuit, 'the ritualized confrontation of the song duel was at the same time an outstanding example of the same principles of indirection, denial of hostility, and pacification that governed nonconfrontational modes of dealing with conflicts' (2000a: 111). These principles include the rule that the opponents were not supposed to talk about the conflict directly but to conceal it with irony and an artistic performance. The distance between the two adversaries, created through the audience, prevented a violent encounter. Hence, in East Greenland just as in the Central Arctic,

> Any resolution of conflict that aimed to keep the accuser and the accused in the community . . . had to deal with the contradiction between the need to

work out conflicts and the needs to avoid a public confrontation or placement of guilt. The song duels provided the means to do this, by creating a safe context and an ambiguous genre for the airing and defusing of conflict.

(Eckert and Newmark 1980: 198)

Similar principles of avoiding overt confrontations or conflicts are still highly relevant to many informal sanctions that point to the values and expectations of the community, as I will illustrate below.

Regarded as 'heathen practices' by the Christian missionaries, the song duels were prohibited in the beginning of the 20th century. Gessain (1969: 122) reports that by 1935, they were no longer taking place. In the 1940s a colonial council was set up for the districts Ammassalik and Ittoqqortoormiit that had the authority to prosecute violation and to pass ordinances. The hunters' association also dealt with juridical tasks during that time (Smidt and Smidt 1975: 245). In 1964, a court of justice was introduced in East Greenland as part of the already existing 'Court system of Greenland'. This court is still working today. Following this system, a district council in Tasiilaq consisting of local citizens and the High court in Nuuk exert jurisdiction. This legal framework takes into consideration the particular circumstances of a case and involves community members in decision-making (Smidt and Smidt 1975: 245). The police station in Tasiilaq forms the executive branch (closely interlinked with the West Greenland headquarters), and police officers travel to the district villages (as well as Ittoqqortoormiit) once a year to deal with delinquents. In the villages, during the rest of the year, two village residents hold the position of 'police helpers'.[4]

Thus, the sanctioning role of drum and song duels has nowadays been partly replaced by modern jurisdiction. While this legal system does accommodate some Greenlandic values, it departs from fundamental premises of interpersonal communication and confrontation avoidance that were, and still are, highly relevant in contemporary East Greenlandic society. The underlying values are integral to how people relate to each other on an everyday basis, and they come to the fore in a variety of ways of speaking and informal communicative sanctions, which include ridicule and teasing, gossip, and withdrawal and ostracism. Before elaborating upon these informal social sanctions, I will discuss some theoretical concepts upon which this chapter is based.

Expectations, values, and social sanctions

I think we have always overgeneralized about the cultural 'traits' shared by the members of a group, but now it is even more irresponsible than it was formerly to make general statements about what 'Inuit' do, think, and feel.

(Briggs 2000a: 114)

This statement by Briggs effectively summarises the problems in trying to write about Inuit values (Briggs 1979, 1982), which do not exist as a uniform system of precepts, and, in fact, never have. This is even more crucial in modern East

Greenlandic society, which has been undergoing rapid social, cultural, and economic changes within a very short time. Accordingly, one encounters an intermingling of various lifestyles, goals, and values, and much disagreement about the legitimacy of social expectations (cf. Briggs 2000a: 115). A closer look at people's practices and communicative encounters, nevertheless, reveals certain core values, as I will show below. These values are constantly renegotiated and aligned to changing circumstances, and are not always shared by all members of society.

It has long been assumed in anthropology and sociology that all societies have ways that incentivise people to conform to collective values and expectations, and conversely that penalise non-conformity, which can be subsumed under the term 'social sanctions'. Where formal, institutionalised sanctions cannot be identified, such as in hunter-gatherer societies without strong centralised authority and evident stratification, it is supposed that sanctions operate informally (Ferraro 2006: 321–3). Accordingly, in East Greenland, informal social sanctions are said to exert a strong influence upon inhabitants' daily lives, which, among other things, relates to the formerly egalitarian political structure. For East Greenland shortly after its colonisation, Holm reports

> the existence of an unwritten social code, the authority of which is tacitly acknowledged, while its transgression entails the penalty of social dishonour. In many respects, indeed, this code will be found to place far severer restrictions on the liberty of individuals than the written law of civilized communities.
>
> (1911: 57)

Somewhat similar to Holm's conception of 'code', scholars would often search for some kind of legal system when studying Inuit societies, and would speak about 'Inuit traditional law' (Pauktuutit 2006: 9) or 'Inuit customary law' (Laugrand et al. 1999a: 1). Drum and song duels have been compared to a 'court of justice' and regarded as an expression of law (Smidt and Smidt 1975; Thalbitzer 1941: 626). Others, trying to avoid a direct invocation of 'law', have made use of terms like 'rules' (e.g., Therrien 1997: 253), 'norms', or 'morals', the latter often conceived of as a body of non-authorised expectations set in contrast to law (Fikentscher 2008: 9; Moore 2003). However, I doubt the usefulness of the notion of law in trying to understand the Inuit practices covered by this concept, and I concur that, as Laugrand et al. propose, terms such as rules and morals 'suggest a much more formalised structure than actually existed in Inuit society' (1999a: 1).

The use, in a number of East Greenland ethnographies, of such terms as law, rules, or morals should not be (mis)understood to imply a static and determining framework, which does not allow for individual expression and agency. Just as Therrien has stated with respect to Canadian Inuit, 'individuals who do not conform to the moral code are not necessarily forced to return within the boundary of normality' (2008: 254, my translation). The strong value of personal autonomy among the Inuit, both in Canada and in East Greenland, entails the right to decide whether to act according to other people's expectations or not; nobody would tell another person what to do, though every person has to bear the consequences of his or her behaviour (cf. Gessain 1969: 41). Hence, if people do not conform to shared anticipations, they might not be directly called to account; nevertheless, they might fall victim to gossip, sarcasm, and social exclusion, or 'social dishonour', as Holm writes above, either temporarily or in the long term. A high value placed on personal autonomy, thus, does not stand at odds with shared expectations of proper conduct, and the corresponding sanctions, despite their subtlety, seem to be well understood by most East Greenlanders. Accordingly, as Thalbitzer wrote of East Greenland during the first part of the 20th century:

> Only occasionally . . . did the individual venture to complain of his neighbour's violation of the customs, e. g. if he violated the rules of taboo or the boundary for honesty and uprightness. In cases where indignation accumulated, the offended one or the guardian of the morals often preferred to expend his wrath in circulating an evil rumour by means of scandal, or by secret persecution with the aid of magic means. Only in cases of open scandal or hostility the offended person had recourse to open persecution before the national court of assize, viz., by the juridical drum-singing.
>
> (1941: 626)

Non-interference and practices of confrontation avoidance

As in all societies, Iivit may not always agree with other people's practices. However, straightforward criticism and giving orders is rare, and I have seldom experienced East Greenlanders approaching somebody in a face-to-face manner, telling him or her what to do. Verbally explicit criticism, talk about confrontational matters, and, more broadly, the managing of information about other people is frequently characterised by some type of avoidance. Conflicts, or even curiosity, might come to the fore in a number of ways, many of which imply indirect ways of speaking and looking that protect another person's autonomy. Confrontational themes may be framed in terms of humour or gossip, or take place by sending a text messages via the mobile phone, with the distance between the communicants facilitating confrontational exchange. Although various institutionalised contexts of modern daily life have brought about changes to direct communicative exchange, which is nowadays perhaps more widespread than before, 'traditional' values of non-interference and not giving orders and criticism still fundamentally underpin most daily social interactions and society at large.

In order to illustrate how this value of non-interference affects ways of sanctioning other people, I want to give a short example of one of the rather exceptional cases of a verbally explicit confrontation that I experienced during my time in East Greenland. One afternoon in June 2007, I was standing next to the post office in Tasiilaq, in the company of my elderly friend Paula from Sermiligaaq and another woman from Tasiilaq. A drunken man was standing next to us; he was singing loudly and moving in an uncoordinated fashion. My two companions and I were bothered, and we exchanged meaningful facial expressions. After a while, Paula uttered, *'tusarngaraalit!'* (you are noisy!), whilst turning her head in the man's direction. He did not react and continued singing. I seldom encountered such straightforward criticism; nevertheless, I found that elderly people often state opinions more frankly than others (which is in line with the grandmothers' influential position within the family context). Yet, my friend's remark also implied some confrontation avoidance. It was framed as a statement that did not urge the man to reply, supported by the fact that she was standing diagonally behind him, and he did not look at her. She did not follow up her criticism, and neither did the people surrounding us. How to react was ultimately up to the man (*nammeq*).

This confrontation avoidance substantially changes under the influence of alcohol. East Greenlanders with a high intake of alcohol quite frequently speak about emotional themes, personal problems, tensions, and rivalries, issues that are usually held back under normal, more sober circumstances (cf. Gessain 1969: ch. 11). Likewise, violence in East Greenland is quite frequently connected to alcohol consumption, which 'acts on the individual as a stimulant of aggression, which is normally contained and controlled' (Robert-Lamblin 1984: 118; cf. e.g. Larsen 1992b, I will return to this theme further below).

Curiosity, approaching others, and keeping informed

The informal social sanctions that I experienced in East Greenland build on particular ways of acquiring information about other people and on particular conventions of approaching others. Most Sermilingaarmeeq keep up-to-date about other inhabitants' daily rhythms, their regular whereabouts, and people with whom they spend time. Due to the small size of the village, most villagers regularly meet each other in everyday life. People are attentive, observant, and regularly ask and talk about other villagers. For example, when staying outside in the village, it is common to ask residents passing by where they are going (*sumut*?), or where they are coming from (*suminngaanii?*). (Sometimes suggestions are made to the person passing by, as for example, 'Are you going to Thomasine's house?') These questions are posed to any person regardless of their degree of familiarity. The latter might then mention a specific place, such as *'pisiniarpimut'* (I am going to the shop), *'uangittemut'* (I am going to my house), or give a vague indication of direction, through a gesture, a nod with the head, or a short verbal indicator, such as *'igani'* (over there); fellow villagers are often able to guess what is meant by it. Sometimes inhabitants also mention a relative's kin term, or the name of the person they are heading to, once again added with the addition of the prefix *-kunni/-kinni*, denoting the whole family group. A person visited is often of the same gender, at least if he or she is not related, or the boy- or girlfriend of the visiting person. It is not common to follow up on questions such as *'sumut?'* or *'sumingaanii?'*, and to insist on further information. In most cases, it seems to be up to the approached individual to decide if he or she wants to volunteer any further information or not. Through such brief questions curiosity is expressed, though ultimately the addressee holds the choice of how much he or she wants to tell. This careful way of questioning and showing interest is also evident with regard to people's travel plans. The first question posed to a person who has newly arrived in the village is usually '*qarnga tigippit*?' (when did you arrive?), and, if villagers expect a person to know a departure time, this is followed up by asking '*qarungu attartsaulit*?' (when will you leave?).

Such news not only livens up people's everyday village lives, but might also be useful or necessary, in one way or another. Nevertheless, as motivations and planning are not always communicated, and since to continue to ask somebody about plans amounts to an improper intrusion, people often extract the information they need from various indirect channels of information. For instance, if somebody has particular interest in knowing about other people's travel plans, he or she will attend to all kinds of signs that could point to preparations for departure. This especially counts for boat travel, since travel by helicopter and supply ship follows a schedule; it is more predictable, and passengers are often known in advance. Accordingly, when I was sometimes searching for a lift in a boat to another settlement, it was best to hang around in the centre of the village to investigate when people were leaving and if there was space in the boat; villagers are sometimes willing to take along others, usually in return for some payment of fuel. I would thus wait for people to declare their travel plans, or for situations which would allow me to pose questions. Decisions, such as whether to give somebody a lift are, nevertheless, 'up to the person' (*nammeq*), and nobody is obliged or expected to give an explanation for deciding against it. I have illustrated this with an anecdote in Chapter Three, in which a young man had departed without giving me the opportunity to ask him for a lift.

Keeping informed about other people's travel plans could also be useful in order to find somebody who might take along packages for relatives or friends at the destination. This is a common practice throughout the region. People from Sermiligaaq, for example, often send bags or parcels with *kalaalimernit* to relatives in Tasiilaq. Transported items may range from foodstuffs to various kinds of goods, and on one occasion I was even asked to take a hamster with me in the helicopter to Sermiligaaq. Such deliveries are often organised spontaneously, for instance by approaching a neighbour who, most obviously, is about to depart, or calling somebody on the telephone when hearing through other villagers about the person's departure. It is thus important to stay alert and to react quickly.

This way of approaching others underlines the situated character of questions and conversations, and relates to the aforementioned fact that verbal inquiry without apparent context amounts to an intrusion. Asking somebody where he or she is going, or if he or she might be able to take a delivery, usually happens when a person is walking through the village, when he or she is preparing to go somewhere, or after the person has told other people about it. Such apparent indication legitimates questioning. This observant attitude and the situatedness of conversations might necessitate quick reactions in promising moments, and the Iivit are often remarkably fast in this respect. For instance, I was amazed by the sudden speed and fitness of a female friend with whom I had been hanging out on the sofa for some hours – watching TV or the like, and not moving much – how quickly she went into action when something important happened down at the pier, such as her husband or some other relatives returning home, or if plans had come up to join a boat trip. As I have described with respect to keeping appointments and being on time, East Greenlanders move, or take action, often very quickly, at auspicious moments when everything comes together. Just as Mazullo and Ingold have described for the Sámi, it is a matter of timing:

> It is not time that is crucial so much as the *timing* . . . what counts is not punctuality but *readiness*, not the precise targeting of a point in time but a continual monitoring of the way things are going, in a world in which everyone and everything is in movement, each at their own pace, along alternately converging and diverging paths
>
> (2008: 34).

There is thus a balance between being attentive, keeping an eye open for the happenings in the village, listening out for what other people have observed or heard, and careful face-to-face questioning. This carefulness in approach is based on a combination of protecting personal autonomy and depending on others – a combination that underwrites the element of trust, as mentioned in the last chapter. For example, practices of filling free boat space with additional people, or of taking along deliveries for other people, depend on the willingness of boat driver, who is not pressured nor generally expected to do so. For these practices to endure, people trust that some of their fellow community members might act according to their wishes, though not all of them will do so. As Ingold has explained, 'Although you depend on a favourable response, that response comes entirely on the initiative and volition of the other party' (2000: 69–70). Trying to impose obligations or to lay down conditions that the other is bound to follow would represent a betrayal of trust, and would interfere with the value of personal autonomy.

The high value placed on personal autonomy thus influences the fact that Tunumeeq often hold back with judging other inhabitants' actions and practices. Nevertheless, though people hardly reproach each other in a direct verbal way, criticism or disapproval might be expressed in a number of indirect ways.

Face-to-face encounters: ridicule and teasing

One way of expressing criticism or dissonance, and of pointing out shared expectations, is ridicule and teasing, which may range from light-hearted joking to derision and sarcasm. One can distinguish two forms: first, teasing (or shaming), usually happening in the other person's presence, thus providing the possibility of laughing together *with* the person in question, and second, laughing *about* a person in his or her absence. In East Greenland, the first variant is particularly widespread, though people also laugh about absent people, sometimes in the course of humorous gossip.

In the previous chapters I have provided various examples of teasing and ridicule, such as of joking about hunters with a record of poor performance or about persons who get lost during boat trips, or of teasing as part of the socialization of children. These examples have illustrated that teasing is an indirect way of communicating disapproval and particular values. Through this way of speaking, responsibility is turned over to the addressee who has to decide how she wants to understand such ironic comments and how he or she wants to react to the wider audience. As Eisenberg writes, 'teasing "works" because of its inherent ambiguity. The recipient must decide whether the speaker is serious or whether she is "only joking"' (1986: 186; cf. Schieffelin 1986). Teasing and irony thus provide 'safe' ways to express otherwise confrontational issues, which do not interfere with a person's autonomy. Just as with drum and song duels in bygone times, the reactions of the audience are crucial, influenced by both the addresser's performance and the addressee's reaction. Humour, moreover, forces one to cope, and 'having to laugh at one's own trials or listen to others laugh at them coerces one into accepting them, foregoing gloom, resentment and anger' (Howe and Sherzer 1986: 688). Accordingly, teasing may intensify the relationship between those involved, which parallels my description of drum and song duels. Laughter and ridicule about disapproved behaviour keeps people in place; it preserves equality by defeating social distance, though at other times it may also sustain inequalities.[5]

In Sermiligaaq, teasing and joking are an indispensable part of everyday village life. At all the meeting places in the village, conversations often include laughter and irony, be it about a person joining a gathering, children playing in the vicinity, somebody walking by, or an absent person featuring a story. Prominent themes include outer appearances, somebody's funny movements, bodily functions, and sexuality. Now and again disapproval is also expressed in this way. For instance, when people gather outside, it sometimes happened that a villager would subtly make a critical comment about somebody else smoking a cigarette, which would result in laughter or smirking among the other people in the group. The person might say, '*sigaritseq mamanngiraq*' (cigarettes do not taste), accompanied by an exaggeratedly distasting expression upon his or her face, whereupon the addresser might reply, '*mamakkaaju*' (they taste very good). Both parties would proclaim their opinions without explicitly interfering in the other's business, even though the first comment evidently implies some criticism of the smoker's habit. Being framed in a funny way lessens the confrontative momentum.

Children are frequently the butt of jokes, as mentioned before, which I will illustrate with an example from a community celebration in Sermiligaaq. Every year on St. Lucia day, December 13, a religious celebration and Scandinavian tradition, Sermilingaarmeeq organise a celebration in the schoolhouse, comprising a procession by a group of girls, wearing white dresses and each holding a candle. During the St. Lucia celebration in 2006, the girl's procession through the basement of the school building was followed by a small boy, wearing a grey anorak with the hood pulled over his head. In contrast to the girls' white dresses this looked rather funny, and, as boys are generally not involved in the performance, the little boy caused much laughter in the audience; yet nobody told him off, and he continued all the way. Hence, laughter and teasing in these examples pointed out expectations and disapproved behaviour, without explicitly telling the person laughed about what to do.

A humorous game played at a number of community festivities in Sermiligaaq holds quite similar functions. The game involves a number of small pieces of paper with short statements written on them by some *inuusuttut* (young people). Some of these statements that I was able to understand were funny remarks and insinuations about particular villagers. The folded pieces of paper were randomly distributed among the audience, and nobody knew which person had written which remark. One by one, each person read out loud the paper she or he had picked, which was met with laughter and amusement in the audience. The anonymity of the writer and the humour involved lessened the statement's confrontational effect. Quite similar to drum and song duels in former times, the game highlights villagers' expectations of right conduct.

Apart from pointing out disapproved behaviour with regard to individuals, humour also provides a means to deal with social and postcolonial inequalities, and plays a part in negotiating and sustaining group identities.[6] As Carty and Musharbash have argued, 'Laughter is a boundary thrown up around those laughing, those sharing the joke. Its role in demarcating difference, of collectively identifying against an Other, is as bound to processes of social exclusion as to inclusion' (2008: 214). Accordingly, Sermilingaarmeeq and other East Greenlanders once again laugh about inhabitants of other local settlements (which often takes the form of gossip), and about Greenlanders from other parts of the country; they joke about *qattunat*, be they Danes, tourists, or foreign anthropologists. For instance, in spring 2007 when I was planning a trip to Ittoqqortoormiit, Sermilingaarmeeq enjoyed joking about the people there, especially about their language, which differs slightly from the Ammassalik dialect. Sermilingaarmeeq, too, are sometimes the butt of jokes in other settlements in the region, and East Greenlanders are made fun of by West Greenlanders. The latter jokes usually build upon widespread prejudices and stereotypes among West Greenlanders, rendering East Greenlanders 'uncivilised' and 'backwards' (these negative attitudes have a long history, as has been explained by Karen Langgård 1998). One such joke was told to me during a stay in Nuuk in summer 2008. One evening I visited a nightclub with a group of friends, where I talked to a local man. Speaking to him in Tunumiusut, I mentioned that I had lived on the Greenlandic east coast for a while. He was quite surprised to hear this, and, with a witty expression on his face, he remarked '*Tunummi krokodilirapoq*' (there are crocodiles in East Greenland). We had to laugh, and I teased him about where he had gotten this information. This joke referred to the prevalent rumours about the East Greenland population (e.g., a high level of violence, suicide, and so forth), about which West Greenlanders receive much negative news in the media (Elixhauser 2009b).

I was also frequently teased and laughed about and humour played an important part in the negotiation of my own role and position in the community. For instance, villagers joked about my mistakes in using their language, especially during the initial stages of my fieldwork; they teased me about my funny movements whilst learning fishing and walking on the ice, and joked about some of my bodily features that differed from theirs. Most commonly, such joking happened in my presence. We had much fun together, and laughing together strengthened social ties and friendships. Nevertheless, I was not aware of some of the jokes, especially during the first months of my fieldwork. For example, during an early visit in Sermiligaaq in summer 2006, I was accompanied by a German friend, who had worked at the Red House in Tasiilaq that summer. I introduced her to my Sermiligaaq friends by calling her my '*qimmeq*', which means 'friend', at least according to what I then thought. Yet I had mispronounced the word for friend (*kamma*), and had called her 'dog' instead. This caused great laughter among the Sermilingaarmeeq, which I did not notice at the time. Just several months later, a Sermiligaaq friend explained the incident to me, and we had to laugh about it together. Hence, situations of being laughed *about* during the beginning of my fieldwork were gradually replaced by being able to laugh together *with* my friends (Dwyer and Minnegal 2008). This anecdote illustrates that humour may serve as a marker for a person's integration into other cultural contexts.

Consequentially, some types of joking about me changed over time, as my own movements in the village became more familiar, and were no longer attended to in the same ways. Yet other kinds of jokes changed less, as they depended on the ambiguities of my role as an anthropologist, which I have touched upon in earlier chapters. For instance, my sustained interest in many realms of daily life which was not confined to 'typical' female affairs was sometimes met with scepticism or a lack of understanding, and resulted in frequent teasing. I was teased for acting like a hunter's wife when being interested in hunting activities; I was teased that I had romantic interests in the male hunters when I showed interest in joining the hunting and fishing crews; and I was jokingly put forward as a candidate during the elections of the village council and the hunters' and fishermen's association, a group whose meetings I frequently attended. Some of these comments pointed out deviances from people's anticipations, and sometimes led me to alter some of my practices in order to avoid being the butt of jokes and the centre of collective laughter. I could not adapt to local habits in every respect, however. Teasing and humour also provided a 'safe' way for the villagers to learn about my incentives and interests. For instance, a few times female friends jokingly suggested that I could start an affair with their husband, which I insistently declined. I understood these remarks as a sort of test of whether I could imagine such a thing. Similarly, people sometimes teasingly suggested boyfriends and marriage partners, or that I could build a house in Sermiligaaq. Teasing, thus, sometimes pointed out my own difference and gave me the possibility to explain my own role and the background to my stay in Greenland. Some of these jokes also showed the people's positive attitude towards my presence, such as when I was suggested as a candidate for the local associations. Accordingly, with Howe and Sherzer, I argue that '[w]hen the members of a society deal with anthropologists through humour, they do more than put individuals in their place. They also create that place, by situating an elusive and liminal social category' (1986: 691). Just like the two anthropologists Howe and Sherzer, I am grateful that Iivit friends created such place for me, put up with me, and laughed with as well as at me.

Sexuality and ritual humour

A widespread type of humour among East Greenlanders is sexual humour and jokes about gender relations. Sexuality quite frequently appears in everyday conversations: possible sex partners are jokingly suggested; people are teased about sexual activities and bodily organs; and jokes are made about other people's (possible) sexual performances. Such verbal humour is sometimes accompanied by obscene gestures, and occasionally humorous exchanges consist in gestures and facial expressions only. In this, particular joking relationships may be in evidence, such as between potential marriage partners and persons from alternating generations. Yet, sexual joking is not confined to a particular group of people, though it occurs rather seldom among children. The latter supports Apte's assertion that 'a relatively low occurrence of sexual and scatological humor among children in a culture may suggest casual attitudes towards sex and body waste' (1985: 107).

Sexuality in East Greenland is dealt with rather casually, and it is often talked about in a matter-of-course way. It is not particularly frowned upon to change sexual partners, especially among the unmarried people, and also among married people love affairs are widespread. Many Iivit recognise a clear distinction between a person's long-term relationship and occasional sexual encounters with other people. Linked to the value of personal autonomy, partners are not allowed to show jealousy explicitly (though there is, in fact, a lot of it), or to reprimand their partner openly for being unfaithful. Christian morality, however, runs counter to such practices, and in general, many people voice their opposition to unfaithfulness within marriage. Actual practices, however, differ, which was once explained to me by a friend. She explained that people are free to play around if they are not yet married, also when in a relationship; within marriage attitudes differ: some wives or husbands would accept if their partners are unfaithful once in a while, whilst others would oppose this, which should also be respected.

It becomes clear that sexual relations are characterised by inherent tensions: Christian expectations of faithfulness and negative feelings of jealousy stand in contrast to the anticipation that a person should not show jealousy about a partner's unfaithfulness, as this interferes with his or her personal autonomy. As Ingold has written with regard to hunter-gatherers more broadly, 'The underside of love and trust is jealousy and suspicion – the suspicion that the person for whom one cares may assert their autonomy by caring for another' (2004b: 173). Sexual joking in East Greenland reflects these inherent tensions, and plays a part in the negotiation of respective expectations.[7]

Sexual humour not only forms part of everyday conversations, but also appears in the form of playful games and performances at various festivities, as, for example, community celebrations during occasions such as Christmas or Easter, or festivities organised by local associations. Participants in these humorous games are usually young unmarried villagers, though sometimes married people take part as well. Often, couples of male and female persons are randomly joined, in order to perform some funny task with sexual overtones. Such performances give rise to much enjoyment and laughter on the part of the other villagers, who might exchange witty comment about the players and their performances.

During the 2006 Christmas celebration at the Sermiligaaq school building, one such game involved a few couples dancing together holding an orange between their foreheads. They would try to keep dancing as long as possible without losing the orange. In another game, a number of women, each with a balloon tied onto her lap, sat on chairs and the partners had to burst the balloons by trying to take a seat on the women's laps. At the Christmas feast of the Sermiligaaq fire brigade, a game was played where some women had to move a coin on the inside of their male counterparts' trousers, starting at the left leg and leading over to the other side. This was followed by a game in which the women had to make dancing movements. The men would then imitate their moves, and the winner was chosen according to the volume of audience laughter.

Hence, similar to drum and song duels, the audience takes part in judging a winner, or a winning pair. Often it is not the fulfilment of a particular task that seems crucial to the performance, but the ability to entertain the crowd. Hence, as with storytelling, performance skills are important, which in these examples necessitate being able to handle activities with sexual undertones with a random partner in a light-hearted and entertaining way.

Such skills also form part of humorous games and performances, for which participants dress up in funny ways. One such example took place during a community celebration in Sermiligaaq in 2006. Two Sermilingaarmeeq had dressed up like a married couple. The woman was wearing a white veil, resembling a curtain, and the groom had put on a nice suit with a hat, decorated with a flower (see Figure 6.4). They marched into the assembly room, along with a villager dressed

Figure 6.4 Staged marriage, Sermiligaaq, December 2006

(Photo: Sophie Elixhauser)

up like a priest, and started to imitate a marriage ceremony. The 'priest's' outfit consisted of a black garbage bag. The villagers in the audience were laughing to tears, and even the players could not hold back their laughter. I am not sure why these two people were chosen, both of whom were married to other persons, and whether the particular pairing influenced the audience's laugher. Nevertheless, the staged marriage illustrates an inversion of social roles, which has often been found to be characteristic of performative or 'ritual' humour (Apte 1985: ch. 5).

This reversal of roles is most apparent in the East Greenlandic winter festival *milaartut*, taking place at Epiphany (called *kongepingasit* in East Greenland), which I have discussed in Chapter Five with respect to gift giving and sharing. The 'cross-dressing-mumming tradition' (Kielsen 1996: 123) is performed by a number of people disguised as grotesque figures, and it entails giving presents to the frightening players, and a dancing competition in the evening. I will now mainly talk about the dancing competition during *milaartut*, and the sexual humour involved, relying upon my material from the *milaartut* festival in Sermiligaaq in January 2007.

Milaartut started on the evening of 5th January. Seven adults had dressed up in a grotesque manner, which made them more or less unrecognisable. Villagers guessed at their identities until the next day, and it turned out that, apart from one married person, the *milaartit* were all young unmarried people. Their blackened faces were distorted by tape, and one of the players had white paper (or band-aid) glued onto his or her face. Some had walnuts in their cheeks, making them resemble Greenlandic masks, which were worn on these occasions in former times (Nooter 1975: 162; cf. Gessain 1984; Rosing 1957). Costumes consisted of old clothing, decorated with cardboard, cloth, and other available materials. All of the *milaartit* had stuffed the inside of their clothes, forming hunchbacks, bulges on the arms and legs, oversized bottoms, large breasts, and penises. In one case, a phallus was made from some cardboard and a piece of fur, tied onto the player's clothes (see Kleivan 1960: 100, for a similar account from *milaartut* festivals in West Greenland). The costumes included colourful wigs and hats, sunglasses, and huge gloves; one player was holding an artificial cigar, two of them had guns with them, and one *milaarteq* was carrying a large thermos. Several players were wearing two different shoes (often oversized rubber boots), with the left worn on the right side, an observation which has also been mentioned by Nooter (1975), who gives a detailed account of a *milaartut* festival in Tiilerilaq in 1968. Most *milartit* appeared in pairs, and in some cases their costumes resembled that of their partner's look, such as each would wear one of the other person's shoes. The exaggerated sexual characteristics, differently sized phalli, and exaggerated breasts and bottoms, did not indicate the *milaartit* to be of one gender, and many players carried both male and female characteristics.

The *milaartit* kept silent at all times, and villagers were allowed neither to speak to them, nor to laugh at them. After some of them had walked around the village in the early evening, trying to scare the inhabitants, they assembled in the *katersutarfik* (meeting house) to compete for the prize for the best appearance, dancing style, and behaviour. A great number of village inhabitants were present (see Figure 6.5). One after the other, the players performed their dance to the

Figure 6.5 Milaartut dancing competition, Sermiligaaq, January 2007
(Photo: Sophie Elixhauser)

music, which included, every now and again, obscene bodily movements, horse-play, and sometimes mimed phallic advances towards the audience. The amused spectators laughed wholeheartedly, and the winner was decided upon by the level of applause.

The East Greenlandic *milaartut* tradition, which today is more alive in the rural places of the region, is related to the East Greenlandic *uuajerneq* games, events characterised by oral performances with disguised players, which ceased to be performed in the middle of the 20th century (Kleivan 1960: 160; Nooter 1975: 165; Rosing 1957). It also bears a close resemblance to *milaartut* practices in West Greenland (Nellemann 1960), and as Nooter explains, 'The main points of agreement are the disguise, the use of masks and soot, the use of clothing of the opposite sex, the carrying of sticks, the silent entrance, and the element of play in the identification of the *milârtut*' (1975: 165–6).

The ritual context of *milaartut* shows an inversion of behaviour with respect to gender roles, and an exaggeration of sexual practices in front of a large audience. It thus stands in line with examples of ritual humour from different parts of the world that illustrate a 'reversal of established or common behavioural pattern, associated with sex, social roles, and social status' (Apte 1985: 157). Reversal occurs not only in behaviour but also in physical appearance. Humour thus allows the expression of thoughts and practices that are not usually expressed in front of an audience. This is reminiscent of the relief theory of laughter, associated with Sigmund Freud (1960), who argued that through humour emotional energy is released, and that jokes provide a window into the unconscious. However, this explanation, which has attracted a variety of criticisms, applies only partially to my example, as in East Greenland sexuality is not particularly suppressed by the society, though it is not normally openly displayed in such an exaggerated way. I therefore concur with Apte's argument that

> ritual humor is not qualitatively different from humor in other social interactions. Rather, the rituals provide a special context whereby humorists and participants are both temporarily free from social sanctions for whatever acts they wish to simulate
>
> (1985: 176).

In sum, all my examples on humour and teasing have illustrated that, as Howe and Sherzer have put it, 'humour is not just *about* something but is a way of *dealing with* something' (1986: 691). This something might be rivalries, social tensions, disapproved practices, nosy foreigners, gender relations, or sometimes simply boredom and wanting to have fun.

As I have shown in some previous examples, humour is closely intertwined with gossip. In contrast to teasing, and many other forms of humour, gossip is not directed at a person face-to-face. This difference, however, is one of degree, and there are various forms of 'humorous gossip and gossipy humour' (Morreal 1994: 47).

Talking about people in their presence

One such communicative mode, which is popular in East Greenland, is a way of speaking about somebody using the third person pronoun, though this person is standing right next to the speaker or within earshot. This way of 'speaking for another', as Deborah Schiffrin (1993) has called it, often happened when people met who did not know each other well. For example, in my description in Chapter Five of the trip to Tiilerilaaq that I took with my friend Hansigne, I have shown that when a person first arrives at a place, inhabitants who are not familiar with this person usually do not ask for identification but make enquiries through other villagers. Such enquiries might take place in the presence of the person talked about (though he or she is interactionally absent), or in his or her absence. Here, I will focus on the former.

One summer's day in 2007, I was standing in front of the shop in Sermiligaaq, together with Maline's sister and her mother Paula. Some people had arrived from Tasiilaq, and a man joined us whom I knew only by sight. Maline's sister and Paula chatted with him. At one point, Maline's sister asked her mother who was the father of this man. Whilst doing so, the man stood right beside her, yet she did not address him but asked her mother about him. Paula told her the name of the father. Thereafter, the man was included once again in the conversation. I have come across many similar cases, and found myself frequently in a situation in which others asked my companions about me in my presence, not looking at me and speaking about me using the third person pronoun. In the last chapter, I have given one such example, in which a man from Kulusuk visited the home of my friend Dorthea, and had enquired information about me.

In East Greenland, 'speaking for another' accords with local forms of politeness, and appears to be a sign of respect and tact, quite contrary to its impolite connotations in some other cultural contexts (Schiffrin 1993). Moreover, as Schiffrin (1993: 236) asserts, 'speaking for another during conversation can have not only local interactive meaning, but broader implications about one's own (and other's) rights, privileges, and responsibilities'. In East Greenland, this communicative mode allows a person to follow his or her curiosity without confronting somebody in a face-to-face manner, in line with other confrontation avoidance strategies. The popularity of this practice, I believe, relates to the fact that a face-to-face approach and questioning may appear intrusive and can create obligations for the approached person to communicate particular information which she might not want to share. Asking others *about* a person, though the person in question is within earshot, seems to circumvent such obligation. Responsibility and obligation for speaking is both transferred and transformed: it is transferred to the person who provides an answer; and it is transformed since the answering person is only able to pass on information that has previously been volunteered by the person in question, or that he/she or others have obtained through observation or through own experience. Speculations are rare, and only verbal statements that are already circulating are passed on. This mode of speaking, therefore, does not interfere with a person's right more or less to determine (*nammeq*) when and what kind of information he or she wants to provide.[8]

In 'speaking for another', the presence of the person talked about is not acknowledged, and the person is almost treated as if he or she was not there. Not only is the third-person pronoun used in speaking about a person, but also the speaker does not look at him or her, and might exclude him or her from the conversation through averted body stances. Such a way of speaking has also been described by Don Handelman who, in an article about a workshop in Jerusalem, subsumes this communicative mode under the label of gossip:

> Modes of attempting to exclude the member gossiped about included directionality of address and eye gaze, use of personal pronouns (i.e. referring to the person gossiped about in the third person), loudness, pitch and tone of voice, and stance or body.
>
> (1973: 213)

As in several of the examples presented in previous chapters, speaking for another illustrates the importance of peripheral vision, which, combined with asking somebody else about the person in question, seems more appropriate than asking and looking at somebody in a face-to-face manner. 'Speaking for another' covers many different themes, and as with teasing, it may also provide a way to articulate personal views and criticisms. Yet praise is also often formulated in this way, though positive remarks are also frequently addressed to a person in a face-to-face manner. Through the communicative exclusion of the person talked about, curiosity may be expressed and values may be pointed out, whilst avoiding confrontational or intrusive situations. This way of speaking shows the close alignment of the protection of personal autonomy with its embeddedness in a wider societal context characterised by various verbal and nonverbal channels of information management.

Gossip: talking about absent people

Other ways of making people adhere to shared values are different forms of gossip, which provide important means to pass on information and to learn about community expectations. With Handelman, gossip can be defined as 'informal communication passed about by persons, through the medium of encounters, which if accomplished will be deemed by members of the setting to be of interest and value' (1973: 211), a definition which also covers practices of 'speaking for another'. Yet, some authors stress the negative evaluation of communicative exchange as a major characteristic of gossip, as for example Jörg Bergmann, who calls gossip a 'morally contaminated' activity (1993: 58). This assumption is again criticised by others who question 'whether any form of talk is ever devoid of moral evaluation' (Besnier 1996: 544). In my account, I will give examples of gossip as an evaluative way of communicating about other persons, called *oraasersivivoq* in Tunumiusut, whilst being aware that it is sometimes difficult to discern the moral tenor of the evaluation. Though most of the examples in the following consist of cases of negative talk about others, gossip, I argue, may also include positive ways of speaking about others.

Gossip operates indirectly, which often implies the absence of the person talked about. Yet, as the above example of 'speaking for another' has shown, in some cases, the person in question might well be physically present but interactionally absent. Thus it would be better to say that gossip does not take place in face-to-face encounters (Gluckman 1968: 32). This way of talking is most effective in dense social networks, and it is particularly feared in small communities, as has been demonstrated in various ethnographic studies (e.g., Brison 1992; Haviland 1977). Karen Brison explains:

> Gossip is most likely to cause problems when individuals are dependent on a small group of people for companionship and aid; in this situation, no one can afford a bad reputation within his or her group since this can bring social and economic disaster. Conversely, where people can easily find help and

friendship elsewhere, they will be less concerned with what their neighbors think and so will often ignore malicious gossip.

(1992: 12)

Gossip plays a part in the negotiation of group membership, and it may entail processes of inclusion as well as exclusion. Gossip is further connected to trust, which is a condition for getting to know about people's gossiping activities (Bergmann 1993: 151). Accordingly, I did not know much about such activities during the initial stages of my fieldwork. Yet as my fieldwork proceeded, and with my growing integration and language skills, I came to know more and more about the importance of gossip: not only with regard to what people would gossip (or had previously gossiped) about me, but also with regard to inhabitants' gossip about one another (cf. Dwyer and Minnegal 2008). People increasingly trusted that I knew how to understand, deal with, and participate in the gossip activity. Nonetheless, I am aware that the presence of an anthropologist is never innocent, and that my accounts of gossip are inevitably influenced by my own presence and experiences (cf. Wilson 1974: 100).

As mentioned above, Sermilingaarmeeq and many other East Greenlanders regularly talk about their fellow inhabitants, about who has attended the religious service, who was seen drinking alcohol the night before, who joined the gambling circles and was most successful, and about all sorts of issues related to romantic or sexual relationships, to name just a few examples. Time and again disapproved behaviour is highlighted, often in rather subtle ways. For instance, people might turn up their noses at heavy drinkers or frequent gamblers, or they might show disapproval that the church, once again, was rather empty on a Sunday, as many villagers had taken advantage of the nice weather and left for a family trip. In view of that, Bergmann has called gossip 'discreet indiscretion', asserting that, 'gossip producers, as a rule, express their disapproval and moral indignation after their gossip stories' (1993: 122). East Greenlanders, nevertheless, are often very careful with judgemental comments, and prefer to pass on a criticising statement coined by another person, without noticeably adding much of their own judgment (though, of course, merely repeating an utterance is far from value-free). Disapproval, nevertheless, is often communicated subtly through intonation, breaks, and length of utterance, or through depreciative facial expressions, gestures, and disapproving sounds.

To illustrate how gossip may point out people's expectations and values, I will give some examples of gossip that I experienced in East Greenland. One winter's day in Sermiligaaq in 2007, I met a friend at the village store, and we went for a walk around the village. Passing by the house of a particular resident, we came to talk about the man's former girlfriend; they had split up only some weeks ago. My friend told me that, according to the ex-girlfriend, the man had been '*naalanngik-kaaju*' (cheeky/not listening). My friend did not add any further explication, or suggest any further details. In many similar situations, I have encountered such brief mention of what other people had said, without further speculation. Sometimes, when two inhabitants would speak about a third person, and one of them asked why this person had acted in particular ways, the addresser would either pass on an explanation by the person in question, cite another person's remark, or state that the person had not said anything about it. I believe that this reluctance to advance personal interpretations relates to the aforementioned idea among the Inuit that one cannot predict or know the motivations of other people, and should not try to assume them (Morrow 1990; cf. Schieffelin 1986, on the Kaluli). The power of words also seems to play a role in this respect, and the impacts of saying something out loud. Accordingly, Tunumeeq often hold back with explicit value judgments, and prefer being able to refer to another person, or at least, more generally, to what 'the other people' say. This carefulness in verbally expressing value judgments is also manifested linguistically, in that East Greenlanders do not swear much, and there are few swear words in Tunumiusut (cf. Rink 1875: 32; Therrien 2008: 272–4, for a similar account from Inuit in Canada). Strong judgemental accounts seem to run counter to a widespread reservation to impose one's own wishes upon, and openly accuse, others. My account from East Greenland parallels the following statement by Therrien:

> As Inuit culture does not present itself as a judgemental culture, involving scales of rigid values, it places emphasis on the quality of rapport to others which might be threatened by concealment.
>
> (2008: 269–70, my translation)

The widespread practice of referencing a gossip story, much in line with the people's carefulness in explicitly expressing value judgments, will be illustrated with some further examples. During my fieldwork in 2006–7, a young local woman was dating a Danish workman. A friend from Sermiligaaq told me that the woman had become rather self-centred and overconfident due to her new boyfriend, to which he added '*iivit orarpoq*' (as the people say). Moreover, one evening in spring 2007, I stayed at a friend and her husband's place. I told them that I had spent the previous evening at another person's place, and that a particular lady from the village had been visiting. My friend asked me if this lady had been speaking much. Thinking about it, I admitted that she had. Her husband then added that many villagers did not appreciate the woman's company. She talks too much, '*iivit orarpoq*' (the people say). Hence, my friend and her husband seemed to support the view that the woman spoke too much, though they did not explicitly say so, and first asked for my opinion through posing a rhetorical question. Rhetorical questions imply some ambiguity and seem to facilitate the expression of personal views. Face-saving functions apply, both to the addresser, who does not explicitly have to speak out his or her conviction, and to the addressee, who may decide him- or herself how to react. The latter has been similarly explained by Schieffelin:

> Unlike negative directives, rhetorical questions put the ball back in the addresses's court, providing him or her with the option to respond (in a limited number of ways) or to remain silent. Rhetorical questions call for no

answer, but they do keep communication open and acknowledge the other. A rhetorical question . . . provides the addressee with some face-saving protection . . .

(1986: 180)

My examples have illustrated that a gossip story is frequently introduced or ended with the phrase 'as the people say', which has been noted for gossip in different cultural contexts (Stewart and Strathern 2004: 55). Both by referring to what other people have said and by making use of rhetorical questions, the performances of fellow community members may be subtly scolded, but the responsibility is deflected to another person or to the group as a whole. Hence, with Karen Brison, I argue that '[g]ossiping is another way in which people can influence others without being held accountable' (1992: 26). It provides a way to verbalise opinions about others, without being held responsible for possible intrusions upon a person's autonomy, just as was the case with drum and song duels, which involved a wider group of people carrying responsibility for judging 'right' conduct. Referring to Canadian Inuit, Briggs explains that,

> when the community is perceived as critical some of the burden of responsibility for criticism is removed from the individual accuser. In this way, the confrontation is diluted and so made safer: for the accuser, who has the backing of the community; for the accused, who need not fear retribution at the hand of a single irate individual; and for the community, which can more easily than an angry antagonist, reincorporate offenders and so reconstitute itself.

(2000a: 120)

Apart from cases of gossip, in which East Greenlanders repeat other people's value judgements, now and then people also quite explicitly state personal opinions, such as in the form of a negative (or positive) experience with, or opinion about, somebody. For example, I have often heard people praising others for being very nice (*kajiminnara*) or skilful (*pikongra*), and sometimes inhabitants also complain that a fellow resident is '*ajerpoq*' (not good), that somebody is ugly looking (*peqqingeraq*), or not very pleasant company (*kajiminnanngilaq*). Here, speakers have to take into account that others might pass on their remarks and reference them.

One incident, in which disapproval about somebody's behaviour was verbally explicated, happened to me one day in February 2007. In the afternoon, some women from the village had come to the home of a friend of mine in order to play cards. I joined in for a while, and after initially losing a few rounds, I managed to win some games. During my lucky streak, one of the participating women mumbled a comment about a 'German bank', which, at first, I did not quite understand. After the gamblers had left, I stayed on at my friend's place, and, a little later, the husband declared that this woman had been '*ajerpoq*' (not good). I asked him why, and he explained that she had talked about me in a derogatory way, calling me a 'German bank', with plenty of money. I assume that this remark was triggered by some jealousy over my luck, building upon the fact that some villagers, who did not know me well, suspected that a person from Germany must have plenty of money. My female friend and her husband did not like the woman's way of speaking, which my friend supported by telling that this woman had also created some mess in the toilet and that the other participants had complained about the smells. Moreover, she had been drinking alcohol beforehand, my friend continued with a disapproving expression on her face, as some participating woman had told her. Hence, my friend's husband's explicit remark was substantiated through citing other participants' complaints, which, to some extent, deflected responsibility for the value judgment.

This example shows that gossip may also avert conflict (Besnier 1996: 546). Due to widespread patterns of confrontation avoidance, neither my friends nor I would have approached this person about her performance, and the complaint about her in her absence served as an outlet for disapproval and anger. Moreover, the woman's derogatory comment about me supported the fact that she was herself later gossiped about, which shows that there are certain limits on how much gossiping a person can do. If a gossiper has gone beyond acceptable limits, the table can be turned on her and she can become an object of gossip herself (Pauktuutit 2006: 10–11). It becomes clear that, as Gluckman emphasised, gossip 'usually consists of comments on the actions of others assessed against codes of values, morals, skills, etc.' (1968: 32). Thus, the gambler's drinking of alcohol before joining the gaming session, and her subsequent creation of a smelly toilet, were not accepted as 'right' conduct. Gossip allows people to assess their neighbours and criticise digressions without breaching surface amity and interfering with a person's autonomy.

Apart from reinforcing group expectations, gossip also implies personal self-interest, as several of my examples have shown (Paine 1961). It can be regarded as a form of political action, and hearing gossip tends to influence the way individuals interpret incidents (Brison 1992). For example, the case of the woman who had told others that her former boyfriend had been 'cheeky' (*naalanngikkaaju*), an expression often used for people who misbehave, obviously involved some personal interest, as her comment provided a valve to express anger about his conduct. Roger Abraham explains:

> Gossip must follow certain lines of argument. It makes a statement of approval or condemnation which reiterates the approved behavioural limits of the group. But it is also a tool by which the gossiper exercises personal control over the talked about person, if only because he is licensed to call the person's name.
>
> (1970: 300)

A high intake of alcohol and misdeeds under the influence of alcohol are frowned upon by many Sermilingaarmeeq, and frequently lead to gossip. In this respect it is important to mention that, in East Greenland, as Robert-Lamblin has explained,

> Most drinkers actually drink in order to reach an advanced state of intoxication . . . East Greenlandic society is practically divided into two groups: those

who like alcohol, and who take advantage of any opportunity to get some at any cost, and those who belong to the temperance league (Blue Cross). Often there are drinkers and non-drinkers in the same family.

(1986: 140)

Though Sermiligaaq, as compared to other settlements in the region, has relatively few people who belong to the first group, alcohol abuse is a recurrent theme of everyday conversations. A number of Sermilingaarmeeq have told me that they consider alcohol to be 'bad' ('*immiaq ajerpoq*'). In this, people usually speak about their personal perceptions only, and not about the collectivity. In line with the personal autonomy that is so highly valued, people recognise that different perspectives co-exist, and that these should not be evaluated relative to one another. For instance, during the initial month of my fieldwork, I once discussed the problems of high alcohol consumption with my friend Maline, who told me that she herself does not consume any alcohol; at the same time she mentioned that her mother is very fond of it, without greatly judging her position (later on, when I got to know her better, I found out more about her personal stance). Many inhabitants recognise everybody's right to decide for, or against, wanting to consume alcohol, and to live with the consequences. Just as in many other situations, one does not try to interfere or to persuade a person to change. Yet misdeeds connected to alcohol abuse, such as fights, theft, and other forms of violence take place rather regularly, and they are talked about in the community. The names of the individuals frequently involved are well known to most villagers. These communicative acts may influence the person's reputation, and, as Machin and Lancaster have argued for gossip more generally, they may serve as a means of reprimanding individuals who cannot be confronted (1974: 627).[9]

When talking about violence in relation to alcohol, it is important to add that particular issues, widely acknowledged as severe behavioural misdemeanours, were traditionally not talked about among the Iivit. These include incestuous relationships or the rape of adolescents, which, accordingly, were not socially sanctioned (see Sonne 2003, for an examination of verbal prohibitions in East Greenlandic society; Therrien 2008, for a similar account from Canadian Inuit). I can only guess the reasons for this silence, yet I do believe that here, too, the mute recognition of an individual's autonomy plays a role. Maybe such misdeeds deviate from acceptable behaviour so clearly that merely to call them by name is already too strong a reproach, without leaving the possibility to deflect responsibility for judging other people's practices. Such verbal prohibitions are manifested linguistically, in that often no Tunumiusut words for these misdeeds exist, though today inhabitants might use the corresponding West Greenlandic words. Although this silence is still widespread, East Greenlanders nowadays increasingly see the need to start speaking about themes like rape or incest, and a general attitude of rejection is openly being expressed. Just as Briggs wrote for Canadian Inuit, in East Greenland, more broadly, 'people recognise a much greater need to talk about issues and problems and negotiate solutions' (2000a: 117). These changes are supported by various government-led campaigns that try to bring people to talk, and to realise that in order to do something about such misdeeds they have to be openly addressed, children have to be educated about them, and so forth.

Returning to the theme of gossip, this is also encountered with respect to egotistic or unsocial behaviour. Gossip may be spread about people who are greedy, or who do not share, building on the fact that the amount of presents given or received, shares contributed by visitors and guests, or the amount of *kalaalimernit* offered during a feast are themes of interest to many villagers. I will give an example, which I experienced one day during my visit in Sermiligaaq in summer 2008. I was about to depart to Germany, and an elderly Sermilingaarmeeq said that I could drive with him to the village of Kulusuk, the location of the airport. I did not know this man well, but was happy to take advantage of this possibility. He demanded, however, a disproportionally high price for the trip. I asked my friends for advice, who disapproved, and told me, 'this guy is expensive (*agisoraq*), don't go with him'. A close female friend called him on the telephone for me, and asked him again about the price. When he insisted on the same amount of money, she told him that this is too much and that he should consider that I am 'like a person from here' (*taannaq soortu tamaanermeeq*). I very much appreciated my friend's effort and support, especially since Tunumeeq do not like to interfere in other people's affairs, and I had noticed that my friend, who wanted to help me, did not feel very comfortable in exerting such direct criticism. The man, however, did not change his mind, and I ultimately found another means of transport. Before I left I was still hanging out at the service house and some other outside meeting places for a while. A number of villagers had heard about the incident and asked me about it. They showed their disapproval, and I was told that the man was known for asking a lot of money.

In his account from Ammassalik during the first part of the 20th century, Thalbitzer mentions similar examples of gossip about people who do not share. He writes that the inhabitants

> distinguish between good people, i.e., such as often make presents, and large presents, and bad people, i.e., miserly or suchlike people who do not make presents. In Sermilik they laughed at Singataaq, who was so thrifty that he disliked receiving guests, and preferred to take his meals at night while the others slept, and who actually let his son, just grown up, pay for the tobacco which he lent him during the scarce times of winter. Kooitse and his brothers were praised because they were so open-handed and extravagant that after a lucky bear hunt, when their house was full of meat, they only let a few days pass before they exhausted this in distributions and gifts.
>
> (1941: 641)

I have experienced a number of comparable cases, in which stinginess was met with disapproval or, likewise, generosity was emphasised. These examples show that Tunumeeq also use gossip to reinforce egalitarian relations among themselves.

Up to now I have mainly discussed gossip about particular individuals. Yet gossip (similar to humour) sometimes refers to larger groups of people, and it may involve comments about a group of people, or about a particular person with reference to his or her group membership. Just as in humour, people gossiping acknowledge each other to be members of the same group, and gossip may serve as a technique for keeping outsiders out and for reaffirming the common bond of the group (Gluckman 1963). Accordingly, Iivit sometimes gossip about *qattunat*, whose lifestyle and social and communicative practices differ in many ways from their habits (compare Basso's 1979, account of Western Apaches' caricature of the white man). Conversely, *qattunat* residents also gossip about the Iivit. Moreover, gossip often occurs between the different settlements in the Ammassalik region (cf. Robert-Lamblin 1986: 54). The latter gossip activities usually run along a line dividing the region into the 'northern' villages – Sermiligaaq, Kuummiut, and Kulusuk – and the 'southern' villages – Tiniteqilaq and Isertoq. Tasiilaq does not clearly belong to either division, yet many townspeople look down upon villagers in general, thinking them somewhat backward.

My analysis has shown that gossiping has various different functions: it provides a lease for aggression or frustration that people may feel, it keeps people in place who may not be confronted, it offers possibilities for pursuing personal agendas, it provides entertainment for other members of the group, and it brings about community cohesion. Like teasing and other ways of speaking, the gossiper's performance and the audience's reactions are particularly relevant. Therefore, the gossiper must properly attribute the gossip story, and must be careful in coming up with his or her own (unjustified) judgments. Hence, as Haviland has stated, 'learning to gossip in a group implies . . . learning what constitutes rudeness, and further learning enough about everyone involved to know, at least most of the time, what to say to whom and what not to say' (1977: 7).

It became clear that gossip and humour share many characteristics and have similar functions. Nevertheless, they also differ in some fundamental ways: Teasing and many other forms of ridicule take place in the presence of a person, and allow the teased person to laugh with the speaker. Accordingly, it may sometimes stabilise relations that are subject to conflict. A person gossiped about, in contrast, is not involved in the conversation, and the message is often less ambiguous. The person in question is not provided with the possibility to react directly and to influence the relationship at stake. Another way of sanctioning behaviour, which excludes a person or group of people even more, is withdrawal, i.e. neither speaking with somebody about a person, nor looking at him or her.

Ignoring, silence, and ostracism

Keeping silent, ignoring someone, and avoiding the presence of a particular person are particularly strong sanctions. I have provided several examples of such practices in previous chapters, in which individuals decided to withdraw from those around them and not to interact with (particular) people, whether household members or co-residents more broadly. I have shown that it is the right of every person to decide not to answer a question, if he or she does not feel like it, and that moving away may communicate that not everything is as it ought to be. Hence silence, averted bodily stance or gaze, and moving elsewhere are ways to deal with confrontational situations (Jaworski 1993: 23) which, at the same time, may indicate expectations of 'right' conduct. The ability to overlook or, likewise, to overhear particular people or happenings in one's immediate surroundings is also evident at the various meeting places in Sermiligaaq.

One such case was given to me by Robert Peroni, an Italian resident of Tasiilaq for many years. He told me of a lady in Tasiilaq who had severe problems with substance abuse. Once, after she had been drinking a lot, Robert saw this woman walking along the street totally naked. Greenlanders who were passing by, however, just did not 'see' her, acting as if this woman was not there. He himself, as he told to me, found it more difficult to just overlook this person, to adopt this way of 'not-seeing'. My friend Anni filmed a similar incident for our documentary. A young couple was fooling around whilst walking up the stairs to a house in Sermiligaaq. On top of the stairs stood the guy's elderly uncle, busily fixing his fishing utensils. Though the young people, quite audibly and visibly, came to stand right next to the uncle, he did not pay attention to them, almost as if they were not there. These examples illustrate that, as Robert-Lamblin has described, Iivit do not interfere 'in other people's affairs even when something is happening very close by' (1986: 145), which may entail that others are sometimes literally overlooked.

Likewise, gender relations are regulated by the expectations that one should not pay too much attention to persons from the other gender when randomly meeting outside in the village, especially persons from outside of the family group. There are exceptions, such as when there are specific reasons for having to talk to somebody, in cases of friendships between men and women, and some elderly people are particularly talkative; nevertheless, in various village contexts I found that addressing somebody from the opposite sex in a face-to-face manner was often avoided, and uttering a joke or asking another person about this person, seemed to be more usual.

Social ostracism provides a further reason why people might decide to ignore the presence of another person. Not to be included in conversations and social interactions is an effective social sanction, especially in the contexts of a close-knit village community like Sermiligaaq (cf. Briggs 1970). In former times this sometimes led even to physical ostracism, i.e. to the expulsion of a person from the community, either temporarily or more permanently (cf. Pauktuutit 2006: 11, on Canadian Inuit). I have not heard of any such cases of physical ostracism during my fieldwork – apart from people who voluntarily decided to leave the community, some of whom were said to have become *qivitteq*, half human and half non-human. I have, however, come across some minor instances of social ostracism, in the context of jealousy and other kinds of interpersonal tensions in which silence or ignoring somebody indicated that something was not as it ought to have been. Even more than sanctions which imply particular modes of speaking, withdrawal implies an avoidance of confrontation. It protects a person's autonomy, in that the confrontative issue is not addressed but the person in question has to decide him- or herself how he or she wants to understand this reaction.

Reasons for withdrawal have been changing over time, and have been compounded by various factors connected to the rapid social, cultural, and economic transformations taking place throughout the 20th and the beginning of the 21st century in East Greenland. Briggs (2001) has similarly explained, referring to Canadian Inuit, that due to an increasingly fragmented world, people, more frequently than formerly, have goals and values that conflict with those of others whom they love. She writes that, 'a feeling of competence and self-direction, a feeling of being in control of one's life, is harder to achieve under present conditions' and, with conflict being frowned upon, sometimes withdrawal is the only acceptable reaction (*ibid*: 243). People find themselves

> not only cross-pressured but cross-evaluated, that is, valued for different and mutually contradictory qualities and behaviours in each context. As a result, it is hard to evaluate *oneself* now. It is difficult to avoid criticism – to be a coherent person in one's own eyes and in the eyes of the world, because the world is fragmented.
>
> (*ibid*: 243–4)

I see many parallels here to the situation in East Greenland, where withdrawal is a widespread reaction to the divergent requirements of modern political and social life, connected to different attitudes and ways of conduct (Danish versus Greenlandic, East versus West Greenlandic, and so forth). These new demands create pressures and tensions; they affect people's self-confidence, and impact feelings of autonomy. As Briggs writes,

> If self-confidence erodes and pressures proliferate, autonomy may become a defence against weak self-esteem, instead of a celebration of strong self-esteem. Such an autonomy risks being expressed in withdrawal, rather than in confident initiatives and socially responsible participation.
>
> (*ibid*: 244)

This might also have influenced the elevated numbers of suicides in East Greenland since the middle of the 20th century that have often been associated with modernization and social change more broadly (e.g. Bjerregaard and Lynge 2006; Hicks 2007; Leineweber 2000; Thorslund 1992). This is a theme, however, with which I cannot deal with further here (for anthropological accounts on the theme see Flora 2012, 2015; Stevenson 2014, on Canadian Inuit).

The public and the private revisited

I have explicated various different modes of communication that allow a person to point out that something is not as it ought to be. These informal social sanctions reveal particular values and individual and shared expectations. The latter have often been subsumed under the term 'public opinion'. Many ethnographic sources emphasise the important role of so-called 'public opinion' in maintaining order in the community, such as Thalbitzer, with reference to his fieldwork in East Green-land at the beginning of the 20th century. After describing the influential role of the *angakkit* (shamans) and the head of a household, he writes:

> What had even greater influence in the maintenance of domestic discipline and reverence for the transmitted customs was 'public opinion,' within the house and settlement. This influence was due to the direction of *ilerqoq*, ('customary use') or, as the East Greenlanders say *par^w ŋutipargut* 'our use and wont,' i.e. the custom of our community, . . . but without the rules being formulated in real laws, by-laws, or even proverbs. No moral code has ever been formulated, but certain myths and tales contain indications thereof.
>
> (Thalbitzer 1941: 625–6)

Various other Greenland ethnographies have used the term 'public opinion', and have described the power that it exerts upon the members of a group (e.g., Gessain 1969: 41; Rink 1875: 32). Likewise, much social science literature identifies 'public opinion' as an influential means of social control (e.g., Ferraro 2006: 323; Price 1989; Phillips Davison 1958). Nevertheless, though East Greenlanders have spoken with me about the ways they do particular things, referring as they do to particular customs, I have not come across a term literally translatable as 'public opinion', and I have doubts about whether this is the right expression to be used in this context. To explain my reservations, I will explore how this expression has been used and defined in academic literature.

In the literature, 'public opinion' is most often only vaguely defined, if defined at all. For W. Phillips Davison (1958), 'public opinion' is a transitory phenom-enon that involves verbalization and communication among many individuals on a widely known issue (cf. Price 1989). He explains that it leads 'to a situation in which the behavior of each member of a public in regard to an issue is conditioned by his expectation that other members of the public hold similar attitudes on the same issue' (Phillips Davison 1958: 91). This notion of public opinion, just as many similar variants, contains at least two problems. The first relates to the term 'opinion', which, as Ingold points out, is often used in the sense of an attribute or belief as internal mental state that is supposed to motivate individual behaviour. Public opinion is often treated as a detachable 'body of ideas', expressed through speech or other communicative modalities (Ingold, personal communication, April 11, 2010). Yet in East Greenland, as in many hunting and gathering societ-ies, individual or shared opinions do not exist apart from the relationships and contexts in which they are embedded, and voice is not regarded as an external expression of inner opinions, attitudes, or desires. Rather, it should be understood as an extension of the person, or as a 'way of being' (cf. Biesele 2004; Ingold 2004b; Strecker 2004). This is closely connected to the power of words in East Greenland, in that words do not just 'carry information' but are perceived as powerful in themselves. Just as Fienup-Riordan has argued with respect to the Yupiit in Alaska, in East Greenland hasty words can be dangerous, which is linked to the aforementioned ideal of staying indifferent and not showing one's anger.

Fienup-Riordan explains: 'Just as the mind is powerful so it is also vulnerable, and hasty words can do great damage. From this reticence to hasty verbalization it followed that value was placed on a person's ability to retain equilibrium in tense situations.' (1986: 264)

The view of speech as an extension of the person is prominent in many hunting and gathering societies. Ingold explains that in these societies,

> The spoken words, through which a person may, in narrative, reenact the events of his of her life, are not detachable. They are not objects that carry the recollection of these events, but belong to the person's very being as it makes its way in the world. Of course persons can – prompted by others' demands – tell of what they know. But this knowledge, likewise, does not exist apart from the knower, as a set of discrete, intellectual products. It is not objectified.
>
> (2004b: 171–2)

This inseparability of knowledge and knower also explains the problem that people in many hunter-gatherer societies have with speaking for another, or with representing others (e.g. Biesele 2004; Philips 2007), which I repeatedly encountered in East Greenland. As I have shown above, Iivit rarely ever talk about other people's opinions or views and, if doing so, then only by quoting people directly. Consequently, what is sometimes called public opinion may, for the people themselves, be a series of interwoven narratives rather than a detached body of beliefs or attitudes. People belong to a conversation and weave themselves into it through histories of mutual relationships. In view of this, social sanctions such as gossip, sarcasm, ridicule, or even ostracism are not so much an expression of collective opinion as indications of conversational breakdown, where personal narratives have, as it were, gone astray and have failed to weave together with the narratives of consociates. Taking this view, one could argue that what are called social sanctions in the literature are actually ways of drawing these narrative loose ends back into the conversation.

The second problem with the notion of 'public opinion' is connected to the meaning of 'public', a term heavily implicated in the discourse of liberal democracy, that rests on a fundamental division between 'public' (between individuals/households) and 'private' (within individuals/households) spheres. This distinction, however, does not hold universally, as has been shown, for example, in anthropological studies of gender.[10] In East Greenland there are many contexts to which the public/private distinction is inapplicable. With regard to informal social sanctions, and particular ways of speaking (and not-speaking) about self and others, along with nonverbal ways of communicating, I was struck by the fact that many themes which, in many other western countries, are often regarded as private (e.g., Inness 1992), are quite openly discussed at various village meeting places. For example, people might discuss such questions as whether a woman has access to contraceptives, has recently had an abortion or is planning one, or whether a woman fell pregnant and by whom. Likewise, they might mention that a particular person is unhappy about her husband, or his wife, being unfaithful, or they might discuss the reasons of the recent split up of a couple. In many cases, personal observations and experiences, and sometimes word-for-word accounts of the person(s) in question, are passed on. Such conversations might involve the person involved, or they may they take place in his or her absence; they may be expressed in humorous ways, in a more neutral mood, or sometimes as gossip. Although practices differ with regard to whether a person is addressed in a face-to-face manner, or whether he or she is talked about in his or her absence, nobody seemed to doubt the fundamental legitimacy of knowing about such issues; they are not treated as particularly personal, and their dissemination does not seem to be restricted to a particular group of people. I had the impression that once a person speaks out about something (or communicates via gesture, facial expression, and so forth), it is available to all. To my knowledge, there are few exceptions to this, as for instance intimate conversations among couples, and in most cases a person speaking about something takes into account the possibility that it might be passed on. This might even involve themes such as emotions, if a person feels like communicating them (though emotions are not often spoken about).[11] Accordingly, I found it impossible to distinguish a particular body of 'private' themes from another corpus that is 'public'.

This conforms with accounts from other small-scale societies, in particular hunting and gathering societies, many of which, as Karen Endicott asserts, 'do not recognize a public sphere versus a domestic sphere' (1999: 414; cf. Ingold 1999b: 407; 1986: 239). Adam Moore (2003), however, takes a contrary view. Defining privacy as 'a right to maintain a certain level of control over the *inner spheres* of personal information and access to one's body, capacities, and powers' (2003: 218, my emphasis), he argues that privacy is a cultural universal, although its particular forms are culturally variable. For him, privacy provides the foundation for any intimate form of cohabitation since, as he explains,

> we cannot imagine a society where friendships, intimacy, and love obtain but where privacy is non-existent. The very relation of association and disassociation that comprises friendship, intimacy, and love is central to the notion of privacy.
>
> (*ibid*: 223)

Moore's argument, however, rests on the flawed premise that where there is friendship, intimacy, and love, there is automatically privacy. His appeal to the 'inner spheres' of personal information, as in the passage cited above, is tied to a Western notion of personhood which posits some kind of 'inner' person, or self, detached from relationships with others in the public domain. This understanding, however, runs counter the East Greenlandic notion of personhood, and likewise, to many other people's lifeworlds, according to which a person is at the outset embedded in relationships with others. Although, just as in every society, one finds friendship, love, and intimacy, in East Greenland there is no self protected by a 'bodily shell', and through their words, people can impact directly on other persons, as I have illustrated in this book. The self cannot be separated from the (social) person, and in most situations one cannot separate out a private realm from a public realm. As I have shown, it is not appropriate to enquire about the unspoken or 'not-communicated' realm, not perceivable by other people, in which people's motivations, thoughts, and feelings often lie. To call this sphere a 'private realm', however – as Briggs does, for example, when she speaks of the 'privacy of thoughts and feelings' among Canadian Inuit and Alaskan Eskimo (2001: 242, 1970: 21) – is to impose an understanding which contradicts the peculiarities of the East Greenlandic notion of the person (cf. Sonne 2003: 204, who stresses the inapplicability of the Western categories of public and private in an East Green-landic context).

It has become clear that, with reference to the East Greenlandic context, although some issues are held back more often than others, using the general label of private in relation to particular themes of conversation does not make sense. Returning to my discussion of informal social sanctions, this runs counter to a widespread view that indirect ways of communication, which characterise many of the above-mentioned informal sanctions, provide a valve to express 'private' thoughts, such as anger or emotions. For instance, in some popular definitions, gossip has been considered as a morally contaminated activity, as it involves the public in discussing private affairs (Bergmann 1993: 58; cf. Wilson 1974; Schoe-man 1994). In East Greenland, however, gossip does *not* entail the dissemination of private themes in public, nor do other informal social sanctions.

Conclusion

My examples have shown various modes of communication that allow people to draw attention to expectations concerning right conduct, whilst not encroach-ing upon another person's autonomy. These communicative practices create and maintain the balance between autonomy and social responsibility that underpins East Greenlandic social relationships at large. All of the informal social sanctions I have discussed entail some kind of confrontation avoidance, a characteristic shared with the drum and song duels of bygone times and which, I argue, is linked to a particular notion of personhood in East Greenland.

Avoidance may be achieved in a number of different ways: in direct face-to-face encounters particular ways of speaking such as ridicule or teasing allow a person to express confrontational themes whilst protecting the other person's autonomy. For instance, humorous communication entails ambiguity in that the responsibility is turned over to the addressee who has to decide how he or she wants to understand an ironic comment and react to the wider audience. When people speak about another person in his or her absence, as for example in the case of gossip, confrontation is avoided in that the person talked about is excluded from the interaction and is not able to influence directly the communi-cative act. Another way of showing that something is not as it ought to be, which excludes a person or group of people even more, is withdrawal, i.e. keeping silent, ignoring somebody, or avoiding their presence. All of the examples show that many Sermilingaarmeeq and other Tunumeeq are very careful in judging other people's practices and that many of their communicative sanctions entail a deflection of responsibility. Informal ways of sanctioning other people's practices are often of a subtle and indirect kind, though they may be very effective and, in most cases, seem to be well understood by the people in question.

These observations are linked to a particular understanding of the person in East Greenland, understood as a relational entity connecting the living and the dead. My analysis leads me to infer that there is no 'inner' self sheltered by a bodily shell, or a 'self image' in the sense of a public person which protects the 'inner' spheres of a person. Accordingly, it makes no sense to distinguish between particular private and public themes, as all issues talked about – or communicated by other means – in some sense seem to be 'public'. Words, or likewise gaze, are dangerous and can directly impinge upon a person, which is one reason for the predominance of indirect forms of communication. This connection between indirect communication, social control, and a particular notion of personhood has been drawn by Fienup-Riordan with respect to the Yupiit in Alaska. She writes that the Yupiit, quite similar to the East Greenlandic Iivit, possessed – and to some extent still possess – a 'highly refined system of indirect, yet none the less effective, interaction, by which a recalcitrant person could be admonished without their mind being injured in the process' (Fienup-Riordan 1986: 264). Hence, as also in East Greenland, in order not to harm the integrity of a person, direct speech and gaze are avoided, and criticism is either held back or expressed through indirect forms of communication (cf. Sonne 2003: 216).

My examples of indirect communication that feature so prominently in people's day-to-day lives and of informal social sanctions respectively, have revealed many similarities with the drum and song duels in former times, in particular with regard to the avoidance of confrontation and the importance of entertaining the audience which is ultimately responsible for judging a case. However, though these implicit and informal modes of communication may at times sanction behaviour effectively, they do not replace the institutionalised context of drum and song duels which brought rivalries and hostilities out in the open. Accordingly, after song contests were abolished, it was 'more difficult to know where an individual or group of people stood, when one no longer had the same means as earlier of learning what they thought and felt' as Kleivan (1971: 32) asserts for the situation in West Greenland. Developments in East Greenland have begun quite similarly, though around one and a half centuries later. With the prohibition of drum and song duels, opponents were no longer able to face up to each other and to express interpersonal or intergroup antagonisms in front of a large audience. Modern (East) Greenlandic society, framed by the Danish model, has adapted only in minor ways to some of the core values of Inuit culture, and has not found a replacement for this institutionalised means of drawing attention to particular values that does not impinge upon people's personal autonomy.

I have explored a variety of informal social sanctions that indicate people's expectations of right conduct, focusing on interpersonal relations among humans. However, in the perspective of many East Greenlanders and other Inuit alike, misbehaviour may not only be sanctioned through other humans, but also through the agency of various non-human entities (cf., Laugrand et al. 1999a, on Canadian Inuit). Social sanctions are thus closely embedded within the respective domains of people's wider relationships with their human and non-human environments, domains which may not be rigidly set apart.

Notes

1 Yet throughout Greenland, and also outside of the country, East Greenland is known as one of the few places where drum dancing and singing is still remembered, and song and drum duel performances on the Greenlandic west coast as well as abroad are often carried out by East Greenlanders.
2 In addition, people would rely on sorcery and witchcraft, and the powers of particular persons such as shamans (*angakkit*) (Robbe 1983; Robert-Lamblin 1996; Thalbitzer 1930).
3 In former times drum singing served a variety of purposes: apart from the songs for the duels, sources report entertainment songs, songs announcing the arrival of a group of people, religious songs, and songs expressing a personal or emotional condition (Petersen and Hauser 2006).
4 The approach of the judicial system means that there is no prison in the country, nor has any such institution ever existed in Greenland. At the moment, dangerous criminals are brought to Denmark, though the building of a prison is much discussed.
5 This argument runs contra Mary Douglas's assertion that joking should be seen as an 'anti-rite' that generally 'destroys hierarchy and order' (1999: 155).
6 Cf. the literature on 'ethnic humour', e.g., Apte (1985: ch. 4, 1987); Basso (1979); and Critchley (2002: ch. 5).
7 In this, I see some parallels to Kazuyoshi Sugawara's (2004) account of possession and equality in gender relations among the |Gui, hunter-gatherers in Southern Africa. Just as among East Greenlanders, for the |Gui sexually-emotionally engaged relationships with another are vulnerable to trespass, and gender relations are characterised by a persistent conflict between accepting the love affairs of others and the 'destructive effect of negative feelings, such as anger and jealousy' (*ibid*: 126).
8 These observations parallel Bambi Schieffelin's (1986) account of the Kaluli, a forest-dwelling people in Papua New Guinea. In the past, the Kaluli did not have formal leadership positions. Just as in East Greenland, the notion of personal autonomy is highly valued, and the power of informal social sanctions is strong. As Schieffelin writes, 'According to Kaluli, one does not know what another thinks or feels. The most obvious manifestation of this is that Kaluli prefer not to talk about or guess what might be in another's mind, but talk about or act on what has been already said or done' (*ibid*: 167). This statement shows close similarities to East Greenland, including the importance of referencing a statement, and the inhabitants' reluctance to speculate about other people's motivations.
9 In the Canadian Arctic people also shame each other on the radio, as Briggs has illustrated in an article on conflict management in a modern Inuit community (2000a). In East Greenland I have not come across a community radio comparable to the Candian case, apart from a programme through which people send personal birthday greetings.
10 Most influential in the early discussions of the distinction between public and private were various contributions to Rosaldo and Lamphere's edited volume *Woman, culture, and society* (1974), followed up by a wide range of studies, such as Landes (1998) and Schoeman (1984).
11 This runs counter to the widespread practice of designating emotions as 'private' at the outset (e.g., Schoeman 1994: 79).

7   The animate environment

Perceptions of non-human beings and the notion of the 'open' person

> When you are out in nature, you become a tiny little human being. The big natural
> forces take over, and you feel great respect for nature. This similarly applies to all
> the spirits and the mysticism abiding in nature. I very much believe in this.
>
> (Aka Høegh, my translation)

This statement by the Greenlandic artist Aka Høegh, taken from an interview in
the Danish newspaper *Kristeligt Dagblad* (28.02.1998, *in* Hindsberger 1999: 106),
describes an attitude towards the environment that I have frequently encountered
among East Greenlanders. According to many, the environment is inhabited by non-
human persons and environmental beings that impact their everyday lives; these
beings have agency, and they can harm or benefit people. The older ethnographic
sources discuss in detail animist beliefs and practices, which are closely connected
to the people's shamanic past. These beliefs still underpin contemporary lifeworlds,
despite the fact that nowadays most East Greenlanders adhere to the Evangelical
Lutheran faith. Yet for many people, the perceived existence of non-human beings
and Christian beliefs in god, the devil, and saints do not greatly contradict each
other, and people will readily draw on one perspective or the other, depending on
the circumstances.[1] The following chapter will show that communicative practices
in East Greenland are closely aligned to the presence of these 'other-than-human
beings' (Hallowell 1955), and communication is deeply embedded in people's relationships with a variety of living beings and entities.[2] Moreover, the Iivit's relations
with non-human beings, and ways of communicating, establish a particular understanding of personhood, based on the notion of the 'open' person.

Definition of animism

Just as Harvey writes of animists in general, East Greenlanders 'recognise that the
world is full of persons, only some of which are human, and that life is always lived
in relationships with others' (2005: xi). Harvey defines persons as beings rather
than objects, as 'volitional, relational, cultural and social beings', who 'demon-
strate intentionality and agency with varying degrees of autonomy and freedom'
(*ibid*: xvii). The notion of animism, which was introduced to anthropology by

Edward Tylor (1913 [1871]) as a 'primitive religion', characterised by a belief that inert things are alive – or that they possess a spirit or a soul – has recently undergone a reassessment (Bird-David 1999). Critics have pointed out that the attribution to animist peoples of a confusion between living beings and objects, between what is alive and what it not *truly* alive, perpetuates a colonial mentality and dualist worldview of Western provenance (Harvey 2005: xii; Bird-David 1999; Willerslev 2007). As Ingold (2006) has argued, animism does not denote a system of beliefs at all but rather a certain ontology of life which repudiates any a priori distinction between the realms of the inanimate and the animate.

If we regard animism as ontological, then categories of living and non-living are not universally applicable, and the term 'person' may apply not only to humans and human-like beings, but also to a far wider community.[3] Animism, Bird-David further argues, should be understood as a relational epistemology, which is 'about knowing the world by focussing primarily on relatedness, from a related point of view within the shifting horizon of the related viewer' (1999: S69). This perspective puts centre stage the question of how persons are to be treated or acted towards (Harvey 2005: xi). It directs attention away from a constructivist view of 'beliefs *about*' to the experiential realities of the people in question, or to the person in his or her 'behavioral environment' (Hallowell 1955).

Stories about *qivitteq* and other non-human beings

During the first few months of my fieldwork in East Greenland, I occasionally heard rumours about frightening non-human beings, but I hardly came across any first-hand accounts. When asking friends or acquaintances about such beings, my questions would often not be taken seriously; people would either tease me about it, or they would state that they do not believe in these things, which belong to the past. But as the months passed, and as I grew closer to people in the community, I began to hear more and more stories about non-human beings, and towards the end of my fieldwork some people spoke to me quite frankly about these beings and would, for instance, warn me of particular dangerous creatures. I learned that many Tunumeeq take the presence and impacts of non-human persons seriously, though attitudes differ, and there are also a number of people who regard this way of thinking as 'superstition' and tell their children not to believe in it.[4]

Reports of meetings with such beings were often characterised by a degree of uncertainty about whether a person had really been dealing with a non-human being, and people would often add words such as *uppa* (maybe) to their accounts. As Oosten and Laugrand have stated with regard to Canadian Inuit, I also found that '[t]oday, when the existence of shamans has become a controversial issue, uncertainty with respect to these encounters is a recurrent feature of many accounts of these meetings' (2004: 87). Similarly, it seemed that the 'doubts expressed also serve to make clear that the one who had this particular experience is no fool' (*ibid*). I shall now give some examples of people's experiences of which I was told, beginning with the figure of the *qivitteq*, a half human and half non-human being.

Figure 7.1 Remnants of Ikkatteq airbase, September 2006
(Photo: Sophie Elixhauser)

There is a deserted military airbase from World War II located alongside the Ikkatteq Fjord in the Ammassalik region (Figure 7.1). The base is bordered by ragged mountains, and nowadays comprises some rusty remnants of buildings and cars, a dusty long runway, and thousands of old, empty or half-empty oil barrels. It was built in 1942 by the U.S. military, and it was used as a stop over for refuelling airplanes during the last few years of the Second World War. In 1959, it was shut down and the area was abandoned (Robert-Lamblin 1986: 87). Up to 800 American soldiers were once stationed at Ikkatteq airbase, and their presence had severe impacts on the lifestyle and culture of the East Greenlandic population (see Figure 7.2). Anthropologists Victor and Robert-Lamblin (1989: 290) called the establishment of the military airbase the 'end of isolation' and a turning point in the history of East Greenland. Denmark's former protection policy could no longer be maintained, and previously implemented restrictions on the import of Western goods were abolished (*ibid*, cf. Elixhauser in press-a).

East Greenlanders and others travelling the north-eastern part of the Ammassalik fjord system – by boat or other means of transport – often pass through Ikkatteq Fjord, which connects the villages of Kuummiut and Sermiligaaq. Sermilingaarmeeq and Kuummermeeq regularly fish and hunt in adjacent areas, and just next to the base there is a creek abundant with arctic char (*kaporniakkat*), which is a popular fishing destination. In the autumn of 2006, I went fishing there with a group of friends from Sermiligaaq and my German friend, Anni, who was visiting for our joint documentary film project. After some hours of fishing, Anni and I walked up to the area of the airbase to take a look around. Our Sermiligaaq companions, however, did not want to accompany us and stayed behind, like many

Figure 7.2 Military jacket from Ikkatteq airbase, Sermiligaaq, January 2007
(Photo: Sophie Elixhauser)

other Tunumeeq who avoid landing at and walking around the area of the base. Amongst other things, this stems from the fact that the former military base is known for its strange atmosphere and is said to be inhabited by *qivitteq*.[5]

Qivitteq are men or women who have left the community and who have disappeared whilst they were out, but who have not died (cf. Robert-Lamblin 1986: 138). They have become half-human, half non-human creatures. Nuttall (1992: 113) explains that the 'root *qivit*- means "disappointed" and the *qivittoq*[6] has become something of a metaphor for the rejection of community.' *Qivitteq* are said to return to the settlements once in a while, mainly during the night, because they need something or in order to harm the inhabitants. The military base in Ikkatteq is often associated with these beings, and when tourist groups visit the remnants of the base, Greenlandic boat drivers or guides sometimes refuse to accompany the foreigners, or will not walk far, taking precautions such as carrying a rifle. I have also heard similar stories relating to a much smaller former military station in the region further south. Likewise, Pedersen (2009b) reports scary stories told about the former airbase in Narsarsuaq, South Greenland, not however involving *qivitteq* but rather some sort of ghosts. According to a rumour, badly injured soldiers from World War II were secretively kept at the Narsarsuaq base's hospital in order not to make them public, and the ghosts of these soldiers are nowadays said to haunt the place.

During my fieldwork I was told about *qivitteq* in various different situations, for example one afternoon in Tasiilaq when the seven-year-old girl Judithe and her brother Mikael came to visit me at the house where I was staying. They were drawing pictures, and Judithe's picture showed some mountains and on top of one of them she had drawn a human being. The girl explained to me that this was a child who had recently disappeared up in the north. The child's mother had been searching for it, but it did not reappear: '*tammarpoq*' (it disappeared). It probably has been taken by *qivitteq*, Judithe went on to explain, and she then added a *qivitteq* in her picture right next to the child, which looked pretty much like a human being (though somewhat bigger). I asked if *qivitteq* always stay in the mountains, and she answered that they are everywhere; they can fly, and they attack children.

A number of Tunumeeq confirmed that *qivitteq* are able to fly and live in the mountains, though once in a while they come to search for people in the settlements. Occasionally, *qivitteq* are also said to inhabit particular buildings such as hospitals. Their presence is not confined to Greenland, for they also appear in Denmark.[7] Petersen (1964: 79) writes (with reference to West Greenland) that *qivitteq* can make themselves invisible and that they are able to speak the animal language; they can read thoughts and their vision is very strong, enabling them to see far distant places. These qualities have also been ascribed to so-called spirits and souls, but in contrast to these, Petersen asserts, a *qivitteq* has a body and is at the same time dead and alive (1964: 85). He explains that a soul must have an earthly habitation, i.e. a body, whereas a spirit does not need a body and also has no need for rebirth. According to him, *qivitteq* do not belong to either of these groups. In contrast to Petersen, Robert-Lamblin subsumes souls under the category of spirits, arguing that 'spirits' in East Greenland encompass so-called 'nature spirits', particular animals, and the souls of the dead (1996: 122). Due to this ambiguity with regard to the definition of the term 'spirit', and because of its connotations of immateriality in modernist dualist thinking (which opposes spirit to matter), I prefer to work with the broader expression 'non-human person'.

But let me return to the *qivitteq*. Another instance of Tunumeeq talking about these beings commenced when I came to stay in Sermiligaaq in July 2006. Just some weeks before, an old man called Ole had disappeared in the mountains. For many weeks the village inhabitants were searching the area around the village, and even a helicopter was sent out at some point. That summer I sometimes took off by myself in order to explore the mountains behind the village, and after arriving back I was asked several times if I had seen Ole during my hike. For several

months the man's disappearance was discussed with search efforts continuing
and people still hoping that he would re-appear. After a while hopes diminished,
and some Sermilingaarmeeq suggested that he might be dead. Remarks about
Ole still appeared once in a while, for example in jokes about my having met him
when walking by myself in the mountains, a habit which the Sermilingaarmeeq
found somewhat strange. Slowly, I began to hear the first rumours about his having become a *qivitteq*. One day in January, the twelve-year-old girl Asta had just
returned to the village after having played with her friends outside in the snowy
hills close to Sermiligaaq. She told me that they had seen huge footsteps over
there, and she suggested that these might have been Ole's. This reminded me of
some stories I had heard about *qivitteq* leaving footprints. The following spring
my friend Hansigne and I were standing outside Hansigne's house, smoking a
cigarette and having a chat. We came to talk about the missing Ole and I asked her
if he might be dead. She answered '*uppa eqqi*' (maybe not). She explained that
he is with the *qivitteq* up in the mountains, and pointed towards the interior of the
fjord. He has met up with two Tiilerilaarmeeq who disappeared a long time ago.
She was convinced that Ole had become a *qivitteq*, just like the Tiilerilaarmeeq.
Back inside the house Hansigne told me that one evening, not too long ago, she
had put up the wet laundry outside her house in order to dry, and the next morning
the clothes were soaking wet again. She was blaming the *qivitteq* for it. A neighbour, she continued, had experienced a *qivitteq* knocking at the door of her house
just two nights before. The fear of frightening beings such as the *qivitteq* confines
Hansigne to her house after darkness, unless in the company of her husband or
some other person. The impact of these beings has been summarised by Robert-
Lamblin. Mirroring my own experiences, she writes:

> Despite assertions to the contrary on the part of the representatives of the
> Lutheran Church, some ancient beliefs have not disappeared, for example
> that wandering spirits threaten and frighten the living, especially when they
> are alone. Curious or disquieting events are often mentioned in conversation,
> suggesting belief in the presence of, or even meetings with very frightening
> supernatural beings. Tales of monstrous-looking spirits are often told and
> taken seriously. There is the *kiiappak* or 'big face' – a mythical being with
> no body, just an enormous head on legs, who often appears at windows dur-
> ing winter darkness and strikes terror into the household; the *kuupaajeeq* – a
> monstrous woman with metallic nails; there are beings that haunt the ice cap,
> and there are the *qivitteq*.
>
> (1986: 138)

Hence, apart from the *qivitteq*, inhabitants experience various other frightening
beings, such as the *kiiappak*, which has been translated as 'mask' and 'ghost'. I
have heard several 'ghost' stories in relation to houses (relating to a number of
different beings), as well as rumours about the expulsion of malevolent beings
and particular people that have special powers in this respect. For example, there
was one house in Tasiilaq that nobody dared to inhabit for many years, due to the

perceived presence of non-human beings. East Greenlanders also talk about certain small human-like people (*iivit miikkaaju*) under the ground and in the interior of the fjords, and little people that help others. From the descriptions I was given, these could correspond to the anthropomorphic 'dwarf spirits' called *iiaayatiq/ iaajivatsiaq* mentioned by Tersis (2008) and Robbe and Dorais (1986), or the 'fire trolls' called *innersuits* that were said to reside under the sea (e.g. Holm 1911: 82; Thalbitzer 1930: 84). Apart from that, I have heard of frightening beings that are said to inhabit the great inland glacier, such as the legendary *timerseq* haunting the ice cap, mentioned by several ethnographers (Holm 1911: 83–4; Thalbitzer 1930: 95).[8]

In former times, shamans were said to be able to see non-human beings that could not be discerned by other people, and some of these beings used to be shamans' helping spirits (*tartat*).[9] Shamans were intermediaries between the human and the non-human world; they had an extraordinary sense of vision, and they could speak with environmental beings and forces. With the help of their *tartaaq*, they would leave their bodies and travel to the 'other' worlds, i.e. the lands up in the sky and under the ground, to compensate for people's misconduct (Robert-Lamblin 1996). Nowadays, people no longer relate the experienced non-human beings to shamanism.[10] When asking about the *angakkit*, I was often told that the last shaman disappeared at the beginning of the 19th century when East Greenlanders converted to Christianity; a few informants mentioned occasional *angakkit* during the second part of the 20th century, and still today, particular people (mostly elders) are sometimes said to have special powers. Nevertheless, though I have sometimes heard a person speculating whether some recently deceased person might have been an *angakkeq* (speculations never corroborated by others), the shaman's functions seem no longer to be exercised. Indeed Robert-Lamblin asserts that, though 'contrary to the assertions of representatives of the Lutheran church, some traditional practices and beliefs have been maintained until today, the shaman's function has definitely disappeared' (1996: 128–9, my translation).

The community of non-human persons experienced by Tunumeeq not only encompasses human-like persons like the *qivitteq*, or beings such as the *kiiappak* or the *timerseq*, but also animals conceived of as persons. The personhood of animals, which I cannot further elaborate upon here, is expressed though a respectful relationship between the hunter and his prey, as has been described in Chapter Three.[11] Furthermore, a number of environmental entities that Western observers would normally consider to be inanimate are regarded, to some degree, as conscious beings with intentionality. For instance, some Tunumeeq say that glaciers (*apuseeq*) are alive and that they may harm people, for example through intentionally moving ice or creating crevices resulting in accidents (cf. Cruikshank 2005, for the perception of 'living glaciers' among the Tlingit in America's far northwest). This was illustrated to me one day in June 2007 when a few friends from Sermiligaaq and I were camping out in the interior of the Sermilik Fjord during a narwhal hunting trip. After we had put up our tents in the evening, huge masses of ice suddenly drifted towards us. The hunters told us to hurry with packing up our tents in order to leave this dangerous area, and get back into the boat, as we

Figure 7.3 Boat trip in Sermiligaaq Fjord, September 2006
(Photo: Sophie Elixhauser)

were almost cut off by a calving glacier blocking our passage. Later on, one of my companions remarked, half jokingly, '*sigeq uumaleq*' (the ice is alive; the ice is a living being). One finds similar remarks about 'living glaciers' in East Greenland ethnographies (Robert-Lamblin 1986: 126; Holm 1911: 79–80).

Another example which illustrates the blurred boundaries between the human and non-human realm comprises narratives about the aurora borealis, which can be found among many arctic peoples (cf. Nelson 1969: 138; Hawkes 1916, on Canada). The corresponding stories in East (and West) Greenland identify the constantly moving northern lights (*arsarneq*) with the deceased up in the sky playing a ball game with the head of a walrus (cf. Nuttall 2008). Johan Uitsatikitseq, Head of the Sermiligaaq School up until 2017, asserted that people in East Greenland used to believe in this story in pre-Christian times (referring to it as 'myth'), pointing out that today a walrus skull is still called *arsaaq*, which also means ball (personal communication, 13 March 2007). Another informant mentioned that the northern lights are the souls of deceased children running after the ball – and not of deceased people in general – which concurs with Holm's account from the end of the 19th century (1911: 83). In official talks, such as in my interview with the teacher Johan, these stories – as with many other non-Christian beliefs – are usually not taken seriously. Nevertheless, I noticed that some people still fear the *arsarneq*, and, in quiet conversations, some East Greenlandic friends would still identify them with the deceased. Unlike the example of the glacier, however, the aurora borealis is not just another environmental phenomenon with consciousness to some degree that happens to be non-human. It is rather a vision of life in the world of the dead. Whereas the glacier is understood as a living being to which people must relate, the northern lights are the reflections of dead children playing. The people thus relate to the children, and/or their souls, and not to the *arsarneq* as such. They are not (like the glaciers) beings in their own right, but rather a manifestation of the beings of dead children.

These stories and fieldwork experiences illustrate that East Greenlanders' life-worlds comprise a variety of human and human-like beings and environmental entities with varying degrees of consciousness and intentionality. These features have often been brought together under the notion of personhood. Thus, these perceptions are closely related to the East Greenlandic notion of the person. Before focussing on contemporary views, I will give some information on the pre-Christian understanding of a person in East Greenland.

The East Greenlandic notion of the person

Pre-Christian understanding

A number of ethnographers have explained that, formerly, all aspects of human life in East Greenland depended on the presence of souls, within and outside of humans. On a broader level, all manifestations of life, all animals, animate entities, and also durable elements of the universe, living or not, had souls. There were no clear-cut boundaries between humans on the one hand and the surrounding environment on the other (Robbe 1981: 75; cf. Ouellette 2002; Therrien 1987, on Canadian Inuit). In former times, a human person (*iik*, pl. *iivit*) was said to consist of a body (*timi*), a double-soul (*tarnit*), and a name-soul (*aleq*) (Gessain 1980; Holm 1911; Robbe 1981; Thalbitzer 1930). According to Robbe, however, these souls (or what ethnographers have translated as souls), 'are difficult to define as they are at the same time a function of the human body, an emanation of mental life, a heritage from the past or several pasts' (1981: 75, my translation). They should thus not be confused with the peculiar meaning of 'soul' that, in the vocabulary of modernism, opposes it to the material body.

In former times it was believed that the double-soul *tarnit* went into a newborn and was expressed through the breath (*anerneq*). This soul was used in its double form (*tarnit*, sg. *tarneq*), because it consisted of two main principal souls, and additionally many little souls which kept the human alive, and which were situated in different parts of the body. After death, *tarnit* became a free soul, and after some period of purification it took up residence in one of the lands of the dead, up in the sky or below the ground. *Tarnit* was bound by the breath but equally by *sila*. *Sila* is the air, universe, exterior, and the weather, but also intelligence, reason, and common sense. *Tarnit* moved the body and provided the basis for physiological life, whereas the essence of a person expressed itself through the name (*aleq*). *Tarnit* (double-soul) and *aleq* (name) were associated, and the breath was regarded as a sign of the presence of this association, appearing at birth only to disappear at death (Robbe 1981; Petersen 1984b). *Aleq* was – and for many people still is – the reflection of all persons who have carried a particular name, and it has often been translated as 'name-soul'. After death, *aleq* remained with the buried body until a child was named after it (Holm 1911: 81). An attribute of the name-soul, as Robbe's informants have stated, is *isima*, which can be translated as thought. Accordingly, the human being (*iik*) was perceived of as composed of a body (*timi*) and of different vital principles such as the double soul (*tarnit*), linked to the breath (*anerneq*), and the name-soul (*aleq*), associated with thought (*isima*) and intelligence (*sila*) (Robbe 1981: 77).

The 'completeness' of the human body, with its different elements, was particularly important. 'The human body formed a whole which got its strength from its integrity' (Robert-Lamblin 1992: 116). Accordingly, people did not cut their hair or nails in order not to risk their health. Likewise, the disappearance of one of the souls (or another so-called life force) endangered the person, and several authors have stressed the importance of 'soul balance' (*ibid*, cf. Petersen 1996: 73; Holm 1911: 81). The different souls were not bound to the body, which was characterised by permeable boundaries, and a soul could sometimes leave it. The double-soul *tarnit* could detach itself from the body, temporarily, such as in a dream,[12] during the shaman's voyage, during illness, or permanently, which provoked death. It could be stolen, which caused sickness, and could be brought back by a shaman with the help of a *tartaaq*. Thalbitzer wrote at the beginning of the 20th century that '[i]f a human is ill this is due to the fact that the *tarnik* [my spelling: *tarniq*, pl. *tarnit*] has been stolen by an enemy, a shaman's irritated helping spirit, or by another offended spirit' (1930: 93, my translation, brackets added). The so-called spirits he names in this respect include, for instance, the inhabitants of the interior, the great inland glacier (Thalbitzer 1930: 93–5). Another evil being that is reported as having 'too much' soul due to its ability to steal souls is the *tupileq* (Petersen 1964: 82, ch. 2). The name-soul (*aleq*) could leave the body when it was offended; for example, if a person voiced the name in an inappropriate situation, the body then fell ill (Holm 1911: 81). As Robbe reports, provoking the name-soul of a deceased before it has taken residence in a newborn child could result in the creation of a malevolent being (1981). In a similar manner, a person's *sila* (intelligence) could be endangered (Rønsager 2002: 30); *qivitteq*, for example, did not have *sila* and were able to steal *sila* from humans, with the help of their 'evil breath' and 'evil thoughts' (Petersen 1996: 71).[13]

People's fear of malevolent beings was thus closely related to a conception of the human person as consisting of a number of souls and life forces. The emergence of various non-human beings was linked to the provocation of a soul and, thus, accompanied any disturbance of the 'soul balance'. Following particular guidelines of proper conduct could prevent this. Both humans (i.e. enemies, see Robbe 1983) and non-human persons could steal souls, which brought illness and, sometimes, even death. In contrast, when a human being had too much soul, he or she would become insane. This was particularly feared, and accordingly, as stated by Robert-Lamblin, mental diseases were considered and treated differently from other illnesses. Such a disease showed of a person too devoted to witchcraft or shamanistic practices, 'that she or he had created in secret a *tupileq*, carrier of evil, which, not having reached its aim, had turned against its creator' (Robert-Lamblin 1992: 120). A person carrying such an illness endangered the group so much that he or she was either abandoned or killed (*ibid*). It is important to add that *qattunat* were not considered to be 'genuine humans' (Sonne 1996; cf. Oosten and Laugrand 2004: 96, on Canadian Inuit). Lacking many of the skills and qualities of the Iivit, they were regarded as similar to children, and their souls were said to be similar to dog-souls (Sonne 1996: 241–5; Holm 1911: 84).

All in all, the presence of souls was not confined to human or human-like beings, but all animals had one or several souls.[14] The animal soul, however, was not the same as a human soul (Thalbitzer 1930: 89; cf. Guemple 1993: 117–8, on Belcher Island Inuit). Likewise, a variety of things and entities, such as particular flowers, graves, or trout streams, were understood to be alive and to have a soul (Petersen 1984b: 633). This understanding also manifests itself linguistically: In her East Greenlandic dictionary, Tersis (2008: 415) states that the word *iik*, which above all stands for a human person, may also denote the 'personification of a natural element' (a formulation reminiscent of the 'old' conception of animism). A related term is *iiva* (*inua* in West Greenland and Canada), which has been translated as 'occupier', or 'the one who occupies' (Robbe 1981: 76). *Iiva* has been described as the principal life force, inherent in all animals and durable elements of the environment. Petersen argued, referring to East Greenland before 1950, that *iiva* (he writes *inua*) is 'a power with consciousness but seemingly without a will; nevertheless it is dangerous to insult or to ignore it' (1984b: 633). According to Robbe, as a life principle for various living and non-living elements of the universe *iiva* did not

Figure 7.4 Graves near Sermiligaaq, September 2008
(Photo: Sophie Elixhauser)

designate life itself. Life was (and still is) expressed through the verb *uumavoq*;[15] *uuma*- denotes all living beings (*uumaleq*) as well as animals (*uumasut*). According to Robbe, the Iivit in East Greenland thus made a distinction between the physical life, i.e. the mechanisms which permitted life (*uuma*-), and the interior life, which allowed an individual to participate in the order of things (*ii*-). With respect to human beings, the latter was expressed through *tarnit* (Robbe 1981: 76).

Contemporary views

Nowadays, people's understandings of the human body have changed, due, among other things, to the spread of the Western health system and the Lutheran church. The absolute principle of the integrity of the body is no longer followed, and the belief in the small specific souls located in various parts and articulations of the body has been abandoned (Robert-Lamblin 1992: 120). The term *tarneq* (or *tarnit*) has been appropriated by the church to denote the Christian soul, and nowadays mostly appears in this meaning. Yet ideas about *aleq*, the name-soul, are still alive. Children are still named after deceased persons, and thereby take on characteristics of the namesake and, in some ways, become this person. More broadly, all the constituents of the person that I have described above still feature prominently in people's everyday conversations, though Tunumeeq have not spoken to me about how they see the specific connections between these elements. *Isima* appears in a variety of terms used in everyday conversations, such as *isimangikkaju* which denotes a wise person or *isimagitsivoq* which means acting thoughtlessly. *Sila* describes a person's intellect – as well as the weather and the world more broadly – and, according to the people, a person who has *sila* acts with reason and prudence, whereas a person lacking *sila* (*silaarngippoq*) has lost his or her reason, is crazy, or mentally disabled. These observations and ways of speaking highlight the importance of what ethnographers formerly called a 'soul balance', i.e. the importance of not losing or provoking *sila*, *isima,* or *aleq*. This still seems to form an important part of human personhood.

When speaking about human personhood, I refer above all to the category of the Iivit. The term Iivit, however, may carry slightly different meanings, according to context. It is mainly used for members of one's own dialect group, though sometimes it is enlarged to encompass Inuit elsewhere or human beings at large. Most often, as Oosten and Laugrand have stated for Canadian Inuit, *inuk* – or *iik* in East Greenland – means 'person, owner, inhabitant', and 'does not refer to human beings as a universal category' (2004: 86). Hence, in the narrow sense of the word, *qattunat* are not regarded as Iivit, though a few are able to integrate by adapting to Inuit values and attitudes. In most situations, Iivit is a 'specific term for people living from the land, sharing a hunting lifestyle and the ideas and values entailed by it' (*ibid*). Though today most Tunumeeq no longer hunt on an everyday basis, many of the values and ideas linked to their former hunting (and gathering) lifestyle remain in evidence, as I have shown throughout this book. The Iivit are aware that *qattunat* share many human features with them, they can have sexual intercourse with them, and there are mixed marriages. Nevertheless, *qattunat* do not have the same relationship to the land as the Iivit and, in many ways, they do not follow Inuit values. Hence, to be Iivit, one must learn to behave properly and to follow certain expectations of 'right' conduct. These expectations have also been mentioned by Fienup-Riordan (1986), writing about personhood among the Yupiit in Alaska, which encompasses humans as well as animals, and which corresponds in many ways to the East Greenlandic notion of the person. Proper behaviour, according to Fienup-Riordan, includes respecting the power of thoughts, words, and gaze, as well as the fact that the mind is vulnerable.

Though people did not speak to me about particular souls in relation to animals and environmental entities, some animals, such as seals, polar bears, and other hunted sea mammals, still seem to be understood as persons – that is, as beings with consciousness and intentions that have to be treated with respect (cf. Chapter Three). Moreover, and as illustrated above, many inhabitants still experience the presence of particular environmental entities with some degree of consciousness or intentionality,[16] though I do not know to what extent these are still understood as having *iiva*, or some other type of soul. Likewise, I can only hypothesise that people's fear of non-human beings is still linked to perceptions about the danger of being robbed of one's soul, or some other bodily component. This understanding was still encountered at the end of the 20th century, and Robbe reported in the 1980s that, according to some Tunumeeq, certain powers of shamans had been transferred to nurses, doctors, or catechists, and that the confidence the people had in shamans was attributed to all persons capable of acting in fields of physical or spiritual health (1983: 31).

In sum, personhood in East Greenland may nowadays encompass a variety of beings, both human and non-human. People were, and to some extent still are, understood as being made up of several so-called souls and life principles, which differ from one being to another. For means of protection against malevolent beings, and to secure the integrity of one's own person, inhabitants had to resort to manifold rituals and behavioural guidelines, which also affected how they communicated amongst each other. Though some of these are no longer followed today, others are still relevant and underpin people's everyday practices at large. After briefly mentioning some such measures reported from bygone times, I will focus in more detail on those practices that relate to people's modalities of communication, both formerly and today.

Behavioural guidelines: human company and the power of speech

> If a human being wanted to live without fear he had to be careful all the time, being cautious of his words and deeds, as there were many spirits which could take offense . . . To prevent this, the Inuit had taboo rules, magic prayers and songs, amulets, and, first and foremost, their *angakkut*. They were major people of influence on the social life, health and prosperity of their fellow human beings.
>
> (Jakobsen 1999: 46)

This passage summarises some of the measures people took in pre-Christian times in order to protect themselves from malevolent beings. Many of these included the help of an *angakkeq* or a sorcerer (*ilisiitseq*), and means such as amulets and magical charms (Robbe 1983; Robert-Lamblin 1996).[17] Similarly, as Robert-Lamblin (1992) explains, particular 'recipes' would keep a person in good health and protect the person against malevolent influence: it was important to consume amounts of food,[18] to dress properly, to know the dangers of the environment and protect oneself from accidents, to follow specific taboos and 'rules of life' during certain circumstances, to guard oneself against evil actions from others (for example, through amulets and charms), and to look for harmony in the relations with the animals and the so-called supernatural world, often with the help of rituals and magic formulae (1992: 117–8; cf. Therrien and Qumaq 1995, for similar 'recipes' among Canadian Inuit). Sources report that garments could protect against evil influence, and that inversions, such as turning clothing inside out or wearing the clothes of the opposite sex, fostered a good relationship with the non-human world (Buijs 2004: 86). There were also various hunting rituals that would appease particular souls or non-human beings, such as the practice of pouring fresh water into the mouth of a caught seal before cutting it up (Petersen and Hauser 2006: 18–9, for other hunting rituals; see Kleivan and Sonne 1985). Such practices, as Thalbitzer explains, were 'due to piety towards the soul of the dead animal and its community with the other animals of the sea (1941: 640). Likewise, people had to refrain from certain actions and ways of speaking when there was a death and during other transitional phases of life, during which their souls were particularly vulnerable (Petersen 1984b: 633–4).

These observances thus not only included specific rituals, such as with respect to hunting or clothing, but also particular ways of speaking and not-speaking, and communication more broadly. Though to my knowledge, many of these rituals are no longer practised today – a cessation linked to the demise of shamanism more broadly – and though I do not know about contemporary uses of amulets and charms, some practices still do come to the fore in contemporary life, in one way or another. I will now focus on one aspect of this continuity: how the experienced presence of non-human beings affects people's communicative practices.

Perceptions of solitude

Attitudes towards being by oneself and being in the company of other people most fundamentally affect, and form part of, communication. During my fieldwork, I found that Tunumeeq did not like being without the company of other people. For example, many inhabitants avoid going into the wilderness by themselves due to the feared presence of *qivitteq* and other non-human persons; if a person repeatedly walks away from the settlement, people suspect that there might be something wrong with him or her, or at least that something secretive is going on, such as an illicit love affair. Accordingly, villagers did not understand my habit of sometimes going walking alone in the mountains, and I was repeatedly warned of the possibility of meeting a *qivitteq* during my hikes.[19] For the same reason, people said that I should be wary of leaving the company of my companions whilst camping out in the fjords during hunting trips. Likewise, many inhabitants of the Ammassalik region are reluctant to walk up the big inland glacier, and foreign expedition groups trying to cross the ice cap to undertake a glacier tour often have problems finding local guides. I have heard a number of times that people fear evil non-human beings in the interior. As mentioned above, particular places away from human settlements, such as the abandoned Ikkatteq airbase, are associated with *qivitteq* (and other frightening beings), and are thus avoided. Hence, quite like the situation in West Greenland, expectations of unknown forces are raised in 'the (demonic) wilderness, and the inland and unfamiliar places' (Pedersen 2009a: 8). As Nuttall has explained, 'Specific places are associated with stories and sightings of *qivittut* [and other non-human beings] and, because there are large areas of the landscape where nobody ventures, these places are imagined to be their most likely haunts' (1992: 113, brackets added; cf. Oosten and Laugrand 2004: 87, on Canada).

Nevertheless, *qivitteq* and other non-human beings also sometimes appear in the settlements and impact people's daily movements therein, especially during the night. Accordingly, some Sermilingaarmeeq avoid walking through the village after dark unless in the company of others. Houses may also be visited by malevolent beings, and are sometimes said to be haunted (remember the 'ghost' house mentioned above), and staying by oneself in a house (*paarsivoq*), especially

Figure 7.5 The mountains close to Sermiligaaq, September 2006
(Photo: Sophie Elixhauser)

when it is dark, is regarded as frightening. Though once in a while one encounters a person living all by him- or herself – usually an elderly person – these individuals are often pitied. For this reason, many Sermilingaarmeeq did not understand why, during the initial weeks of my stay in the village, I had chosen to stay all by myself at a workers' hut (*paraki*) next to the school building, which I had been able to rent from a private company in Tasiilaq. Friends would recurrently invite me to stay over at their places and, after I had moved in with my friend Maline's family, I felt that this was much more acceptable. Many inhabitants further avoid sleeping by themselves in a room, and prefer to share a bedroom with others (be it family members or hunting partners), even if another empty room is available. Though the Western demand for spatial privacy – such as the 'fashion' that children need separate bedrooms, is more and more in evidence and put into practice, especially in the towns – I met a number of Tunumeeq who expressed their surprise that anyone should prefer to sleep by him- or herself in a room. Hence, when staying over at East Greenlandic friends' houses, time and again I was offered the opportunity to sleep with all the inhabitants together in one room instead of taking an empty room just for myself (which was also seldom available). Stuckenberger has reported similarly from her fieldwork in the Canadian Arctic (2009: 7). She recounted that in Qikiqtarjuaq, Nunavut, some Inuit did not like their children to sleep in separate bedrooms, and some rooms upstairs were not so much liked as said to be haunted by malevolent beings.

I assume that this fear of *qivitteq* and other non-human beings, which is especially pronounced at particular locations, during the night, and in solitude, could be connected to the reported understanding that a person meeting a *qivitteq* will be taken along and turned into such a being (Petersen 1964: 79). According to oral history, as Sonne explains,

> Integration between spirit and human societies was ruled out. Whether an ordinary human being in the myths is stolen away or chooses by his/her own free will to live and intermarry with spirits, he or she is doomed either to 'go native' (apart from a single visit back home) or return home for good, killing or leaving the spirit spouse and children behind.
>
> (1996: 246–7)

In the literature there are several stories of persons encountering a *qivitteq* – or several such beings – initially assuming that they are dealing with another human person, who thereafter are no longer able to return to the human community (e.g., Petersen 1996: 72). Ole, in the story recounted above, was also said to have met up with other *qivitteq* in the mountains, though I do not know whether these or other encounters were held responsible for Ole's not being able to return to the village.

All in all, situations of being scared (*ersivoq*) or frightened of something (*annilarpoq*), which are often connected to experiences with non-human beings, are quite regularly mentioned in people's conversations, and stories of such situations circulate among the people. These experiences also manifest themselves in dreams (e.g., in *orumangiit*, nightmares), which are often taken seriously. In line with what some ethnographers have remarked with respect to other Inuit around the circumpolar North (see e.g., Bordin 2009; Law and Kirmayer 2005), I found that people's dream realities were considered no less real than the realities of their waking lives. Some conversations with East Greenlandic friends revealed that dreams may connect the dreamer to the invisible world; and just as Bordin found for the Eastern Canadian Arctic, dreams are used for 'communicating with the dead and in particular for the transmission of names from deceased to new born, for receiving messages from other realms of reality, and for various predictions' (2009: 6). This was illustrated to me one day in January 2007 when I visited my friend, Haraldina, in Tasiilaq. She told me that her sister-in-law had recently dreamt about her deceased son, and that this might mean that her son would soon be reborn by her sister-in-law, and then possibly be given back to her. Hence, she referred to the passing on of her deceased son's name (*aleq*) to the future child of her sister-in-law who might possibly give the baby to her for adoption.

Nevertheless, and despite attempts to avoid being exposed to frightening situations, so long as they are in the company of others, East Greenlanders very much enjoy all sorts of spine chillers and horror and ghost stories, be they told as stories, transmitted via the television and radio, or watched on DVD. During my time in East Greenland, I would recurrently sit together in the company of East Greenlandic friends, watching some creepy TV show or listening to a story involving some scary non-human being(s). Most often people would indulge in scare-inducing entertainment in the evenings, under cover of darkness, and in the presence of other people. The activity of watching films or listening to radio shows is framed and bounded, and thus manageable, unlike encounters with *qivitteq* or other frightening beings, which are for real and take people genuinely by surprise. The predilection to 'frighten and amuse' (Nooter 1975: 165), which is widespread among other Inuit as well (see Pedersen 2009a, 2009b, on West Greenland), and which I have illustrated in Chapters Five and Six with respect to the dressing up customs during *milaartut*, is further highlighted in people's penchant for making faces, a practice which was formerly also manifested in the form of masks (cf. Geertsen 1994; Gessain 1984; Rosing 1957).

Clearly, the fear of malevolent beings is tied to particular locations, and especially to being in solitude. Pedersen has remarked: 'The demonic happens in the solitude, where all physical and psychological senses are sharpened, and it is in the solitude [that] the individual is subject to unknown forces . . .' (2009a: 8, brackets added). The presence of other humans seems to provide some protection in this regard, protection – I assume – against losing integral parts of one's own person (be it *isima*, *sila*, or similar components). Accordingly in East Greenland, 'being alone' does not mean 'alone' in the literal sense of the word, since even when there are no other humans around a person may still be surrounded by manifold non-human beings. Without the company of other human people, however, a person is particularly exposed to the possible dangers that these non-human beings bring with them. There is no bodily shell, mask, or 'self image', that could protect a person in this regard, and the mind and other components of a person seem to be easily detachable. Hence, I suggest that due to its permeability and 'openness', the person is particularly vulnerable to the influences of non-human and human powers and has to be careful in his or her actions and ways of communication.

The power of words towards human and non-human beings

This carefulness in communication is particularly evident with respect to the power of the spoken word, which I have already elaborated upon in earlier chapters. For instance, and as mentioned in Chapter Two, in some cases I encountered a reluctance to pronounce personal names, above all of people who have died or are in danger of dying (e.g., a person who has been brought to hospital in a very bad condition) (cf. Gessain 1980: 415); the use of personal names may be circumvented by employing the kin term when addressing or talking about a person. This practice, as stated in the ethnographic literature, is connected to the fear of provoking the name-soul of a deceased person. In former times it was said that such provocation could result in the creation of a malevolent being (Robbe 1981: 52). This example shows that speaking among humans may be closely connected to the existence (or creation) of non-human beings, and that the boundaries between humans and non-human beings are easily crossed.

Specific ways of speaking and not-speaking are thus not only relevant to communication among humans, but also ensure a harmonious relationship with the environment and its various beings. Accordingly, as has been suggested by some linguists' analyses of the East Greenlandic language, the particular power of words is also operative in interactions with the non-human environment (e.g., Dorais 1981: 46). For example, when moving through the vast land- and seascapes of East Greenland, too much speaking is frowned upon, and as Gessain (1969: 140) has written, it is important to respect the 'silence of nature'. This also includes respect for hunted animals, which are thought to give themselves up to the hunter instead of being killed on his instigation. As I have shown in Chapter Three, this respect implies particular communicative norms between the hunter and his prey, such as with regard to speech or gaze. The importance of keeping silent, the power of speaking out personal names, and, quite similarly, the power of gaze were also noted by Holm when describing how people would formerly appease a glacier. He recounts:

> On the southerly East coast, on the other hand, the people have a dread at calling the ill-omened glacier by its right name, Puisortok, especially when they are about to pass it. While the passage of the glacier is taking place they may not, so they say, speak, nor laugh, nor eat, nor use tobacco, nor look at the glacier . . . Above all the word Puisortok ('the place where something emerges') must not be named. The glacier is called instead: Apusinek ('the place where there is snow').
>
> (1911: 79)

Robert-Lamblin writes that according to East Greenlanders at the end of the 19th century, glaciers could listen and were responsive to sounds. People had to approach glaciers without a sound, otherwise the glaciers could get angry with them (1986: 126). She further reports on sacrifices made in order to appease glaciers, which have also been mentioned by Holm (1911: 79–80). This brings to mind my own experience of a 'living' glacier, recounted above, while camping out in the fjords with some friends during a narwhal hunting trip. When the hunters told us to pack our things and to leave the camping site that was about to be blocked by the calving glacier, I had briefly laughed and dropped some joking comment – not yet having understood the seriousness of the situation. My companions censored this laughter, and I was blamed for it a number of times after my return to Sermiligaaq. Connecting this incident to the ethnographic sources that tell of the responsiveness of glaciers and the importance of not offending them with words or laughter, I wondered whether the criticism of my laughter could be related to persisting perceptions that such an act might induce a negative reaction on the part of the glacier. Though I am not sure about this point – and the interpretation could be somewhat far-fetched – I have the impression that some aspects of communication among humans, such as the power of words, or of sounds more broadly, also apply to communicative encounters with non-human persons or also with the souls of dead people.

The latter, for instance, was evident in children's responses to the northern lights (*arsarneq*) that, as explained above, are said to be deceased children in the sky playing football with a walrus skull. During the dark times of the year, I could often observe groups of children whistling at the *arsarneq* – sometimes throwing dog's excrement after them – and running away thereafter. Johan Uitsatikitseq has explained to me that they play with the fear that through their whistling the deceased will come down to get them. One finds comparable stories concerning the northern lights in the Canadian Arctic, and French anthropologist Fabien Pernet has reported similarly for the Inuit in Nunavik. He asserts that a child's whistling and whispering is a form of communication with the so-called spirits (personal communication, 24 May 2010). Likewise, Richard Nelson writes that, according to the Inuit, you should not whistle when you see the aurora borealis since in that case 'dangerous spirits' would come down to get you (1969: 138; cf. Hawkes 1916: 153).

Though often spoken about in a half-serious tone, this children's game and the perceptions surrounding it once again draw attention to the power of sounds and the importance of keeping silent, which features so prominently in communication among humans. This power is linguistically evident through manifold indirect expressions and ways of paraphrasing, as I have shown in previous chapters, and as a number of linguists have contended with regard to other Inuit groups (e.g., Morrow 1990; Therrien 1987, 2008). My examples concerning glaciers and northern lights show that in East Greenland particular ways of speaking and not-speaking (as well as other communicative modalities) are deeply embedded within human-environmental relations more broadly. This has been summarised by the French-Canadian linguist Dorais. He writes:

The elaboration of a descriptive and metaphoric vocabulary, to speak about dead people, animals, . . . and so on, would have been part of a general attitude of cunning and carefulness in dealing with the environment.

(1981: 46)

This power of words, and of sounds more broadly, not only comes to the fore with respect to practices of holding back or refraining from verbal utterances in order not to provoke certain forces, but also with regard to specific uses of the spoken word (see previous chapters). One such area of use is in speaking about illnesses, which is closely linked to what I have already explained about the constituents of persons and about the embeddedness of people's communicative modalities within their wider relationships with the environment.

Speaking about illnesses

The sources report that formerly the wellbeing of a person was intimately bound to the person's relations with, and actions towards, the human and non-human environment (e.g., Robert-Lamblin 1992; Robbe 1983). Accordingly,

> Pain and unhappiness were attributed to a disturbance of the social order and one had to identify a responsibility which was the actual cause of disorder: a breach of taboo, an act of witchcraft, a deviant behavior, an unconfessed fault . . . etc.
>
> (Robert-Lamblin 1992: 118)

Apart from the treatment of minor wounds with specific remedies, such as blubber, many diseases would call for a shaman, who 'looked for the cause of the illness in the past actions of the sick person and of his immediate family', and who would sometimes search for a lost or stolen soul with the help of an auxiliary spirit (*ibid*: 119). This understanding has been reported from different Inuit groups around the circumpolar North, and Therrien and Qumaq have asserted that illness in Inuit societies should not be regarded as an isolable event, independent of human action and environmental powers, but as a process within the enfolding of wider relations (1995: 83–4; cf. Laugrand et al. 2002). Hence, responsibilities for diseases would often be attributed to a human person, who could influence non-human actions. As is mentioned in various ethnographic sources, only the confession of a fault, a bad action, or an act of witchcraft could save the life of a culprit (e.g., Robert-Lamblin 1992: 120; Petersen 1964: 76; Sonne 2003: 204). As Robert-Lamblin explains,

> In traditional Ammassalik society confession was supposed to neutralize the power of things that could otherwise have evil effects – chants, magic formulas or actions. Thus formulas pronounced in front of several people would lose their force.
>
> (1986: 138)

Similarly, in Northeast Canada, a sick person 'would have to say, "This is why I am sick. This is what I have done." Then the person would start getting better' (Tungilik, *in* Laugrand et al. 2002: 40). Emphasising the role of non-human beings in this respect, Oosten and Laugrand 'infer that especially the non-social beings observe or hear confessions' (2004: 107). It is much the same in East Greenland; here, too, non-human beings are reported to have an extraordinary sense of vision and hearing, and are involved in retaliation for transgressions (see above, and cf. Robert-Lamblin 1992).

Though I do not have detailed information on Tunumeeq's contemporary perceptions of illnesses and the reasons for them, inhabitants' ways of speaking about illnesses and injuries particularly caught my attention throughout my fieldwork. Illnesses would be a recurrent theme of conversation among the Sermilingaarmeeq, and most villagers would be up-to-date on each other's health problems. For example, whilst sitting in the service house or on the benches in front of the shop, people would speak frankly about all kinds of physical issues, such as the loss or gain of body weight, defecation, illnesses, and small wounds and injuries, of which people expected others to be able to give the exact story of how they had come about. These issues were of common interest to villagers, and sometimes I was surprised to be questioned by a person whom I hardly knew about some illness or minor injury that I had sustained. Often, indeed, I paid no attention to how and where some scratch or cut had happened to me, a fact which was met with surprise by my interlocutors, who were always able to explain the exact origins of their injuries.

These direct enquiries were all the more remarkable because on so many other topics people would be rather careful with their questioning, at least so long as there was no prior sign of the willingness on the part of the other person that he or she would want to talk about a particular issue. As I have argued before, this is linked to the high value placed on personal autonomy. Most upfront questions related to some shared context or knowledge, be it a picture on the wall, a person's outer appearance, or some occurrence which both interlocutors had experienced (see Chapter Four). With respect to illnesses, however, such a shared context was not always there and, though the appearance of a person would sometimes indicate that he or she was not well, not all diseases were apparent to others. I thus got the impression that holding back information on the topic of illness is much less accepted than other, less socially relevant topics. This impression is consistent with my understanding of the social nature of illness, and on the importance of 'confessing' – that is, speaking out loud – the fault or bad action that is responsible for it.

It is possible, then, that observed ways of speaking and questioning indicate that still today, a person's actions may be held accountable for illness. In the early 1990s, for instance, Robert-Lamblin experienced the following:

> In intimate conversations, supernatural or magical causes are still being given as explanations for illness or unhappiness occurring without obvious cause. A sudden disease, an accidental death while hunting, a mysterious

disappearance or the sterility of a woman are still often attributed to acts of witchcraft perpetrated in secret by certain people wanting to hurt others by pronouncing certain words, or casting spells by creating a *tupileq*. It is also attributed to the intervention of supernatural beings, of wandering spirits which threaten and frighten the living, generally when they are alone.

(1992: 120; cf. Robbe 1983)

Though I do not know whether and to what extent witchcraft is still held responsible for illnesses or other negative occurrences, whether *tupilit* are still created (some informants said that they are not), or whether magical spells are being used, I have the impression that decisions on whether to speak or not to speak about illness are still of concern to more than the individual in question.

A further point mentioned in Therrien and Qumaq (1995: 82), and which I was also told a number of times by Tunumeeq, is that to be healthy one needs to be with other people. Illness, according to Taamusi Qumaq, may develop due to the absence of social interaction, and he argues that instances of loneliness, i.e. of young people locking themselves up in a room, lie at the root of many contemporary social problems such as suicides (cf. Flora 2012). On a general level, of course, the importance of human company for being a healthy person could apply to almost any society. Yet, my interpretations of people's perceptions of solitude, and of non-human beings respectively, lead me to conclude that in East Greenland the advice to avoid solitude when one is sad or lonely also relates to the perception that non-human beings are particularly dangerous when one is alone.

The 'open' person

In East Greenland, one can never be fully alone. Even in the absence of human company, one is always surrounded by various non-human persons. These non-human persons are potentially dangerous. It is specifically to secure protection against these possibly malevolent beings that people seek the company of fellow humans. This suggests a particular understanding of the physical body, and of the boundaries between human and non-human persons. As I have explained above, a person is understood as comprising a number of different components, which may sometimes detach themselves from the body. A person is thus inherently vulnerable. There is no bodily shell, mask, or 'self-image' in the sense of a 'public' person which could protect a person's self. As I have argued in earlier chapters, the self and the person are one and the same, and people can directly impinge upon one another, be it through words, breath, thoughts, or gaze.

Accordingly, the mind, intelligence, and other constituents of a person are particularly vulnerable to people's communicative (and other) actions, carried out not only by humans but also by non-human beings. The latter may have excellent eyesight and may be able to read thoughts. *Qivitteq*, for instance, are said to be able to attack a human person with their evil thoughts and evil breath, as explained above. And shamans, who in former times served as the intermediaries between human and non-human worlds, were said to have a very good sense of vision, and were thus able to see non-human beings that could not be discerned by humans. The sources report that with the help of magical power (*iisimala*), shamans as well as some other persons were capable of harming others through thinking (Robbe 1983).

Given this 'open' concept of the person, which is permeable and may easily fall out of balance, people needed, and still need, to be particularly careful in the ways they communicate (and also in their ways of thinking, cf. Fienup-Riordan 2005). This carefulness also applies to communication with (and about) non-human persons and environmental beings, which are aware of humans' actions. Oosten and Laugrand have stated that in Northwest Canada,

> Inuit are convinced that non-human beings and animals observe them and are aware of their actions . . . Non-social beings are therefore often the agents of retaliation when social beings transgress social rules . . . People are all the time visible to those non-social agencies that observe whether they act as social beings.
>
> (2004: 107)

This seems to be quite similar among other Inuit groups as well, such as in East Greenland where illnesses were, and sometimes still seem to be, regarded as the outcomes of improper behaviour, thus affecting the balance of the different constituents of a person.

David Riches (2000), speaking of Inuit more broadly, calls this understanding a 'holistic person discourse'. He argues that narratives 'confirm the importance of a moral life, but, more fundamentally, holistic person discourse related to the basis of this life, founded in the contradiction between egalitarianism and autonomy' (*ibid*: 679). This holistic and 'open' notion of personhood reveals the embeddedness of a person within egalitarian fields of relations with the human and non-human world. Though a person him- or herself is responsible for his or her actions, the latter might influence other persons, and, conversely, he or she might be influenced by the actions of others. People's communicative practices, though based on autonomous decisions, are from the outset aligned to the presence of various human and non-human beings.

Conclusion

Just as Hallowell has argued for the Ojibwa, I have shown that among the Tunumeeq 'the "social" relations of the self [or person] when considered in its total behavioral environment may be far more inclusive that ordinarily conceived' (1955: 92, brackets added). Among the Ojibwa as in East Greenland,

> the self may interact socially with other-than-human selves, so in the moral world of the self the acts for which the self may feel morally responsible may not all be attributed to waking life, nor to a single mundane existence, nor to interpersonal relations with human beings alone.
>
> (*ibid*: 107)

Alene Stairs (1992) has called this an 'ecocentric identity', which encompasses the human, animal, and material world. She opposes the Inuit ecocentric identity, or 'world-image model' of identity, which is based on relations with both the human and the non-human world, with the conventional 'self-image concept' of identity of the Western tradition, which is individualistic in nature and detached from broader environmental relations. Among the Inuit, she argues, 'The person exists only against the ground of community and the non-human world' (*ibid*: 123). Accordingly, in East Greenland, people's lifeworlds are fashioned from interactions with both humans and non-human beings, which take place in waking life or sometimes also in dreams. Human existence depends on respect and responsibility towards the environment and its various beings, an observation which has been reported from other peoples around the circumpolar North as well as hunter-gatherer societies elsewhere. Ingold states:

> Over and over again we encounter the idea that the environment, far from being seen as a passive container for resources that are there in abundance for the taking, is saturated with personal powers of one kind or another. It is alive. And hunter-gatherers, if they are to survive and prosper, have to maintain relationships with these powers, just as they must maintain relationships with other human persons. In many societies, this is expressed by the idea that people have to look after or care for the country in which they live, by ensuring that proper relationships are maintained. This means treating the country, and the animals, and plants that dwell in it, with due consideration and respect, doing all one can to minimise damage and disturbance.
>
> (2000: 66–7)

Of the Koyukon in Alaska, Richard Nelson has written that '[h]umanity acts at the behest of the environment' (1983: 240). Likewise, the Iivit in East Greenland do not attempt to control their environment, but rather concentrate on controlling their relationships with it (cf. Ingold 2000: 72). To achieve this, they need to be particularly careful in the ways they communicate with (and about) other human and non-human beings. This carefulness is closely linked to a particular understanding of the 'open' person, whose boundaries are permeable, and which is especially vulnerable to the influences of human and non-human forces and means, such as speech or gaze.

Notes

1 To some extent, animist and shamanic understandings have also merged, or have been integrated with, Christian conceptions. I cannot further elaborate on this here but see, e.g., Hindsberger (1999) and Oosten and Laugrand (2004: 85, on Canadian Inuit).
2 It is beyond the scope of this book to explore the role and influences of the Christian mission in this regard (see e.g. Hindsberger 1999).
3 See, for example, Viveiros de Castro (1998) on Amerindian perspectivism, which illustrates a very different concept of animate vs. inanimate. Among Amerindian peoples, 'the original common condition of both humans and animals is not animality but rather

humanity' (1998: 472); this point of view defines the subject and not the object, and any being that is the point of reference (subject) defines itself as human.

4 In an article about North Siberia, Rane Willerslev discusses joking about spirits and the non-human world, arguing that the ethnographer should not take 'animism too seriously' when indigenous people hardly do. Underlining the importance of laughing at the spirited world, he shows that people in North Siberia 'are quite serious about not taking the sprits too seriously' (Willerslev 2012: 21). This shows similarities to the East Greenland case, where one encounters much joking about the non-human world, which does not, however, mean that people do not take non-human beings and entities seriously.

5 At that time, however, our companions did not speak about perceptions of *qivitteq* in relation to the base; it was only later that I heard of the stories told, and put this experience into context.

6 *Qivittoq* (pl. *qivittut*) is the West Greenlandic spelling.

7 For stories about *qivitteq* in other regions of Greenland, see e.g. Flora (2015), Nuttall (1992: 112–4), Pedersen (2009a, 2009b), and Pedersen (2005).

8 Further non-human beings mentioned in the literature include, for instance, the *erkiliks* that are said to dwell on the inland ice and are inimical to humans (Holm 1911: 84; cf. Rasmussen 1921), and the *anersat*, translated as 'spirits' and 'ghosts' (Robbe and Dorais 1986).

9 For further information on shamanism in East Greenland, see e.g., Jakobsen (1999), Robert-Lamblin (1996, 1997b), and Thalbitzer (1930).

10 Cf. Ouellette (2002: 124–5) for similar observations about perceptions of *tuurngait* among Canadian Inuit.

11 For further information about animal 'persons', compare Fienup-Riordan (1986) on the Yup'it in Alaska.

12 Cf. Bordin (2009); Law and Kirmayer (2005); and Therrien (2002), on dream experiences in Inuit societies. Similar to Thalbitzer, Bordin writes that 'sleep, dream, disease and death are all linked to one another through the movements of *tarniq*, the shadowy and immortal component of the person, his vital principle' (2009: 4). Yet, due to *tarniq*'s capability to leave the body, '[t]he bond between *tarniq* and the body (*timi*) is . . . not very strong' (*ibid*).

13 I do not have information on whether *qivitteq* were said to have *aleq* or *tarnit*.

14 Dogs' souls, nevertheless, seemed to be eclipsed, or their development retarded, by the overwhelming influence of their human master's soul (see Thalbitzer 1930: 89; cf. Guemple 1993).

15 Therrien (1987: 163) shows that the word stems *innu-* and *umma-* exist in many Inuit languages – the former usually applied to a human being, the latter to anything that is alive, such as an animal. She explains however that the two meanings sometimes overlap: *innu-* may be used for animals and *umma-* for humans, in some dialects more and in others less.

16 I also heard people saying that particular winds have 'animate' characteristics, in particular that the *pilarngaq* (wind blowing from northwest) is female and the *neqqajaaq* (wind blowing from the northeast, bad weather) male (cf. Gessain 1969: 19). Robert-Lamblin reports on sexual differences between heavenly bodies (sun and moon) and the seasons (winter and summer) (1996: 127).

17 Also *tupileq* were sometimes used in the respect (Petersen 1964, 1984b).

18 Sonne explains that according to East Greenlandic oral history, the ability to procure and eat Greenlandic food 'makes a genuine human grow and survive' (1996: 245). Still today eating Greenlandic food is considered a marker for a *qattunat*'s adaptation to Greenlandic ways of life.

19 Nuttall has reported similarly from his fieldwork in Northwest Greenland (1992: 113).

Conclusion
Nammeq and ways of communicating

This book has illustrated different contexts and settings that characterise day-to-day life in Sermiligaaq, and other places in the Ammassalik region in East Greenland. By doing so, I drew attention to the predominance of particular forms of interpersonal communication, many of which are indirect. These subtle and indirect ways of communication emphasise the high value placed on personal autonomy, expressed through the concept *nammeq;* they serve as a means to negotiate the interplay between personal autonomy on the one hand and social responsibility and expectations on the other. All of the chapters of this book have shown that communication is closely aligned to the particular spatial and social settings where it takes place, and to the varying contexts in which it is embedded. People always communicate in certain places, amongst certain people, by a combination of certain means, be it words, gaze, and other sensory and bodily means, or via material things such as gifts or food. East Greenlandic communicative practices, accordingly, differ according to the contexts in which they take place, whilst revealing major consistencies across the board.

During my fieldwork, the expression *nammeq* was frequently used by East Greenlanders to point out the value of personal autonomy. *Nammeq* means that a person has to decide him- or herself, and that something is up to a person. The particular care not to impinge upon another person's autonomy is closely related to the egalitarian ethos in the region. Many relatively egalitarian groups around the world, especially of hunters and gatherers, have been found to appreciate the value of personal autonomy, whilst at the same time practising close and 'immediate' social relationships (Bird-David 1994; Ingold 1999b). While this characterisation applies in large measure to my findings, it is important not to confuse the personal autonomy characterising East Greenland with the individualism as commonly understood in the West, since from the outset this autonomy is embedded within social relations. Thus, the balance between social relations and responsibilities on the one hand, and the protection of other people's personal autonomy on the other, can be regarded as a fundamental tension characterising social life in East Greenland. The carefulness in people's communicative practices and processes of approaching others is manifested in various ways, including ways of speaking and not-speaking, ways of looking, gestures, mimicry, bodily postures and engagements, and certain media such as material items and technologies. With respect to speech, for example, there is a widespread reluctance to give orders and tell others what to do, which prevents the development of pronounced hierarchies. I have illustrated this with regard to the education of children as well as in situations of leader- and followership. All in all, among the Tunumeeq, conflicts are not usually talked about explicitly, and showing one's anger is rare. Likewise, when unable to refer to some shared context or knowledge, people often hold back with questioning others and wait until a person volunteers information or shows a willingness to talk about something. It is fully acceptable not to answer a question, and to keep silent on an issue one does not want to talk about. This appears in various situations of daily life, be it with respect to communication in one's own home, during visits, or during people's encounters at various outside meeting places.

Verbal communication, nevertheless, often entails hints and insinuations that indicate personal opinions, shared values, or deviation from expected behaviour. For example, instead of giving orders, people might suggest a course of action or pose a question. They might speak in a half-serious voice, tease, or make fun of somebody. Here, as with suggestions or questions, the addresser deflects responsibility: it is up to the addressee to decide how seriously he or she wants to take a remark. Confrontational verbal exchange may further be facilitated by certain media or technologies, such as the VHF radio, which provides protection through the distance between the communicants, quite similar to mobile messages. These examples draw attention not only to the importance of protecting other people's autonomy, but also to the power inherent in words. There are no 'private' themes of conversation, as distinct from others that are 'public'; if something is spoken, it is available to all. Words, which may be regarded as extensions of persons, cannot once spoken be unsaid, and the speaker takes responsibility for his or her words, knowing that any statement will possibly be passed on. Accordingly, people are particularly careful in what they say.

Verbal accounts are often accompanied by gestures, mimicry, or bodily postures, and often communication takes place through nonverbal means alone. Yet these communicational modalities are not necessarily any less direct than verbal speech. Gaze, for instance, like words, is considered particularly powerful. In many contexts people learn not to look directly at certain people or happenings in their immediate surroundings (or not to pay attention to them in any other way), even though they are actually able to perceive them. I illustrated this, for instance, in describing ways of showing curiosity and of looking through the crannies of a house, and practices of retreating into yourself whilst being in the company of people. Here, just as in many other situations, peripheral areas of the visual field are particularly important, as they are less intrusive than focal gaze. Apart from averting one's gaze as well as the aforementioned forms of verbal indirection, there are various indirect ways of communicating based on an avoidance of direct face-to-face confrontation. These are usually well understood by the people concerned. For example, *not* saying something or enacting particular bodily postures can be as clear as an utterance. Moreover, it is quite common to avoid the company of particular people or to leave a place in order to avoid confrontational situations, thereby signalling that some issue lies unresolved. The importance of respecting the personal autonomy of others, however, and the carefulness in communication to which it is linked, seems to apply not only to communication among humans, but also, to some extent, to people's communicative encounters with non-human beings. Thus the power of words, and likewise of gaze, also appears in people's interactions with the non-human world, such as in their encounters with certain animals or glaciers. This once again shows the interpenetration of human and non-human worlds in East Greenland

Despite this apparent recognition of the value of personal autonomy expressed through diverse communicational means, East Greenland is marked by a strong sense of social cohesion and mutual responsibility. Shared values and expectations of 'right' conduct, however, are not usually insisted upon explicitly. Instead, they are communicated indirectly though stories; particular ways of speaking such as ridicule, teasing, or gossip; or facial expressions, bodily postures and other types of meaningful behaviour. East Greenlandic society emphasises relatedness, be it among kin or the community more broadly, and values such as hospitality and sharing (the importance of which, however, has decreased considerably). Inhabitants often rely on their social networks – which may include non-human beings such as animals – trusting that others will respond favourably to them. For instance, people travelling to other settlements do not always arrange accommodation beforehand; they trust that their relatives will be willing to host them – or any other inhabitant of the place if there is nobody to whom they are related. This reliance on others is combined with a particular carefulness not to impose expectations on others, which results in various implicit and indirect ways of communication. Nonetheless, sometimes particular services or goods are demanded when it comes to the sharing of food and gifts, or hospitality more broadly. These demands, however, are also often framed in a way which leaves the ultimate decision up to the addressee, providing less of an intrusion into the latter's personal autonomy. Apart from subtleness and indirection, a number of practices at the heart of people's sense of community and social cohesion imply rather unambiguous behavioural signs that are very clear and visible. For instance, in order to enact relatedness, be it among kin or non-kin, and to 'prove' it to the wider community, the value of presents and the amount of meat or other foodstuffs shared is particularly important and is communicated in very direct and explicit ways. This kind of information circulates among inhabitants, who are highly aware of who shows gratitude and acts in a social kind of way and who does not. This influences the status and reputation of people.

In sum, I found a close intertwining of the value of personal autonomy and various social responsibilities, accompanied by certain obligations and expectations. This interplay results in different tensions which characterise Tunumeeq interpersonal encounters and communicative practices at large: tensions, for example, between not interfering with another person's affairs and pointing out shared values and expectations of 'right' conduct, between demanding certain services or goods and waiting for the other person to offer as he or she should, or between staying indifferent to a partner's unfaithfulness and expressing one's jealousy.

Nevertheless, and in addition, the many subtle and indirect forms of communication also tell us something more fundamental about the East Greenlandic notion of the person.

On East Greenlandic personhood

The East Greenlandic understanding of the person, which I have called 'open', is based on the idea that the (human) person is made up of several so-called souls and life forces, including *aleq* (name), *isima* (thought), and *sila* (intelligence), all of which may sometimes detach themselves from the body. The balance of these elements depends on a harmonious relationship with the environment and its various beings, based on the recognition of particular values and ways of proper conduct, some of which affect people's communicative practices, both towards the human and the non-human world. Misbehaviour may result in one of the person's constituents being lost or stolen, which causes illness or sometimes death. Both humans and non-human beings may be held accountable for the detachment of one of these components. Accordingly, a person in East Greenland is a 'partible' and highly permeable entity, which is not bounded by the body (cf. Strathern 1990: 185). Various elements may circulate as parts or extensions of the person, which, apart from the aforementioned souls and life forces, may also include elements such as words or gifts. These extensions do not exist apart from the person in question. The name-soul *aleq*, for example, provides the link between the former person who has carried the name and the newborn it is being given to. Likewise, a present persists as a present *of* somebody, even though it has (officially) changed owner. Words passed on to others, too, are not objectified entities but always refer back to the person who has uttered them. Hence, the integrity of a person is closely connected to proper behaviour, be it with respect to correctly referencing a verbal account, the appreciation of values such as the sharing of gifts, and a variety of other guidelines that prevent the detachment of particular elements of a person.

This 'open' personhood in East Greenland seems particularly vulnerable to the impacts of speech, gaze, and other communicative modalities characterised by a direct face-to-face approach. For instance, and as mentioned above, the power of words, and likewise of gaze, is apparent in many contexts, as it is among other arctic peoples, such as the Yupiit in Alaska for whom hasty words can be dangerous and do great damage (Fienup-Riordan 1986: 264). This leads to circumlocution of various kinds and the manifold subtle and indirect ways of approaching other persons. Accordingly, my observations lead me to me infer that the East Greenlandic person does not comprise an 'inner' self protected by a bodily shell, and a public persona that would shelter the self from the impacts of words or other communicative modalities. Selves or persons – I have used the latter term to denote what others have called both 'person' and 'self' – seem to leak out into the world on the current of words, gaze, or even thoughts, and are thus extremely vulnerable. My East Greenland material thus leads us to question the universality of the common distinction between the self, considered as a psychological substrate, and the person, regarded as the public image of the self and as the social entity which comes into being only when the self enters into social relationships. In contrast to this, a person in East Greenland is relational through and through. The continuity of personhood is provided, amongst other things, through the name-soul (*aleq*) through which a newborn baby takes over characteristics of a deceased person, and in some ways becomes this person. I thus concur with Rosaldo's (1984: 146) argument that 'an analytic framework that equates "self/individual" with such things as spontaneity, genuine feeling, privacy, uniqueness, constancy, the "inner" life, and then opposes these to the "persons or "personae" shaped by mask, role, rule, or context' is based on our own Western assumptions and may prove misleading as a frame to explain other peoples' understandings.

In conjunction with this dichotomy, the person is often conceptualised as a sort of mask that shelters the self from particular impacts from the 'outside'. This mask is sometimes called 'face' (e.g., Brown and Levinson 1987). Yet, with regard to East Greenland, I argue that the person (or 'face') is not a mask that would protect an 'inner' self. There is no 'hiding place' for the self to inhabit, such as a private domain of interaction, and through their words and ways of looking, people can impact directly upon one another. Accordingly, communication may be particularly dangerous, especially those communicational modalities, such as direct speech or gaze, which do not entail circumlocution or avoidance of face-to-face contact. In this context, a division between direct and indirect communication should separate those ways of communication that are potentially dangerous for the person and his or her autonomy from others that are not, instead of being drawn along a line separating verbal from nonverbal modalities.

The examples presented in this book have illustrated that the kind of indirect communication which protects another person's autonomy, as well as personal integrity more broadly, may be achieved in a number of ways: it may take place through verbal communication which is either indirectly formulated or indirectly addressed, or through ambiguous ways of speaking, such as in jokes. It may rest on ambiguities concerning the identity of the speaker, it may involve practices of averting one's gaze, or it may entail meaningful acts such as moving away. In many situations, the ideal that one should make oneself clear and communicate directly, which holds in various Western societies, does not hold for East Greenland. Here, speaking directly about particular things or people and the use of focal gaze is often less socially appropriate than indirect ways of communicating. For instance, just as Rosaldo (1984: 150) describes in her fieldwork among the Ilongots, anger among East Greenlanders appears as a thing which, if made explicit, can destroy relationships. There is no general claim – as in some other societies – that anger should be released in order to be manageable. Moreover, for Tunumeeq the sociality of vision does not lie in direct face-to-face contact but in the operation of peripheral vision which, just like various indirect ways of speaking, is social precisely because it preserves autonomy and protects the integrity of a person.

On a broader level, the inseparability of self and person is linked to an egalitarian ethos and to the relative absence of hierarchy, as has been shown by a number of ethnographic accounts from other hunting and gathering societies (cf.

for example, Ingold 1991; Myers 1986; Rosaldo 1980; Strathern 1990). In relatively egalitarian societies such as in East Greenland, the need to control other people (just as various non-human beings and entities) is at odds with the high value placed on personal autonomy; people's reservations with respect to direct speech – such as in expressing orders – hinders the development of pronounced hierarchies. Rosaldo has summarized the link between this understanding of personhood and the relative absence of hierarchy. She writes that, with respect to the conventional juxtaposition of the self and the person,

> What is not recognized is the possibility that the very problem – how society controls an inner self – may well be limited to those social forms in which a hierarchy of unequal power, privilege, and control in fact creates a world in which the individual *experienced* constraint.
>
> (1984: 148)

Many relatively egalitarian societies, she suggests, do not see the need for such controls, nor do people in such societies experience themselves as having boundaries that must hold in check certain forces or desires in order to maintain their status and engage in everyday cooperation.

Some broader implications

In this book, I have mainly focussed on people's informal interactions and ways of communicating. My work is primarily focussed on interactions and observations in households, outside places, and community buildings in East Greenlandic settlements, and in boats and in campsites out hunting and fishing. I have set to one side the many official and institutionalised fora typical of modern everyday life in East Greenland. These fora are often characterised by various tensions between indigenous Greenlandic values, the Danish, and, more broadly, Western values that have fundamentally shaped the political, economic, and social system in Greenland. These latter values, I argue, are based on a different notion of the person and on different premises of interpersonal communication. They call for direct face-to-face, frank interactions and practices of decision-making that are often at odds with Greenlandic notions of personal autonomy and integrity. Moreover, they are based on a rigid distinction between the human and the non-human worlds, which – though increasingly taken up by East Greenlanders (at least in official talk) – does not correspond with many East Greenlanders' experiential realities. These different underlying premises, I believe, lie at the core of the observable difficulties in recruiting socially responsible participants to the institutions of public and political life. Although these staffing problems have become less prominent as East Greenlanders adapt to modernisation, they are still much in evidence today. One of many examples is the widespread reluctance to launch personal economic initiatives in East Greenland, such as setting up a company. The premises of modern entrepreneurship remain at odds with the disinclination to give orders and to tell others what to do.

The 20th century has seen rapid political, economic, and cultural changes in East Greenland, and with it, an exponential growth of social problems and violent and aggressive behaviour (e.g., Larsen 1992a, 1992b; Robert-Lamblin 1984). Explanations for the increase in violence and aggression range from the broad (i.e. modernisation) to the specific (i.e. the abolition of drum and song duels). Robert-Lamblin writes about an 'internal transformation of social mechanisms that has led to situations where aggressive and violent behaviour is no longer controlled or canalised' (1986: 146), and provides a number of explanations and examples. She includes the disappearance (or weakening) of old mechanisms for reducing inequalities, such as sharing practices, the introduction of wage labour based on individualised principles, less social integration of the individual, the weakening influence of the 'elders' and of 'traditions' (which, I argue, includes changing norms regarding communication and face-to-face interaction), and the fact that peaceful mechanisms for dealing with conflicts such as drum and song duels no longer exist and have not been replaced (*ibid*: 126–7). Regarding the latter, while these duels have indeed played a role in maintaining healthy social relationships, I nonetheless agree with Kleivan that '[w]hether there really was a clear correlation between the abolition of the satirical songs and the alleged increase in violations of the customary norms is uncertain, for so many other factors were involved' (1971: 32). Other factors that have been drawn on to explain the increase in violence include the introduction of alcohol to the Ammassalik region, which was prohibited until 1956, and afterwards – up to today – has been restricted to various degrees (Robert-Lamblin 1984: 117). Alcohol lowers people's inhibitions, allowing them then to express what normally must be concealed, and it serves as a safe valve for tensions, rivalries, and jealousies that lack other channels of release. Alcohol consumption does not, however, explain the increase in violence and aggressive behaviour, but should rather be regarded as one of the outcomes of complex developments and rapid societal changes in East Greenland.

Accordingly, with Sonne (2003: 117), I wonder whether violence could be understood as a sign of the rejection of cultural mores imposed from the outside. If so, this indicates the need to search for different strategies to tackle the manifold social problems in the region, ones that better take into account the modes of thinking and core values of the Iivit. Such strategies to seek out practices of social control that make sense locally could possibly be developed in a similar way to the 'Interviewing Inuit Elders' series on 'Traditional Law' among Canadian Inuit, which was based on workshops held by community members (Laugrand et al. 1999b). These workshops did not address Western perspectives on law, or Western interpretations of 'Inuit law', but took as their starting point Inuit concepts that used to provide order in the communities and from there sought ways in which to adapt these concepts to the context of modern society. Just as Briggs has asserted for Canadian Inuit, nowadays also for the Tunumeeq, 'Social order has to be renegotiated on a new basis and new modes of communication more appropriate to life in a large and diverse community have to be found. Most importantly, those new modes have to involve *some* confrontation' (2000a: 116–17). Yet, despite changing contexts and circumstances and the need for some adaptation to the contemporary social, political, and economic institutions in Greenland, I suggest that such strategies should draw on, and further develop, already existing, localised approaches that are based on ways of communicating that allow individuals to address societal and interpersonal tensions without interfering with the autonomy of others.

This book has shown that East Greenlanders constantly negotiate the tension between respecting the personal autonomy of others and acting in a socially responsible manner. This personal autonomy, expressed through the concept of *nammeq*, can be regarded as one of the core values of East Greenlandic society. In contrast to studies that presuppose a sharp distinction between 'traditional' and 'modern' values in Greenlandic society, I find a marked continuity of values in East Greenland, values that are sometimes called 'traditional', many of which are underscored by an egalitarian ethos that is still widespread today. The various social, cultural, and economic changes that have affected East Greenland throughout the last century have provided new contexts and modes of expression for these values, but they have only slightly altered their significance. I therefore argue that in order to better understand the East Greenlandic Iivit, the importance of respecting *nammeq*, and of protecting other people's personal autonomy and integrity, have to be taken into account. The many subtle and indirect ways of communicating which derive therefrom should not be dismissed as 'traditional' practices that belong to the past or that ought to be replaced by 'modern' ways of communicating. Rather, there is a need to better integrate these values into the workings of social, political, and economic institutions that currently govern a 'modern' East Greenland. This could result in more successful communication between Iivit and Danish residents (including some West Greenlanders), or foreigners more broadly, and thus help provide a framework for more positive social development.

Glossary of East Greenlandic terms

aleq:	name
alivarpik:	school
ammassat:	small capelins
anerneq:	breath
angakkeq (pl. angakkit):	shaman
arangi:	tomorrow
arsarneq:	northern lights
iik (pl. iivit):	human person
iiva:	life force inherent in animals and environmental entities
ilisiitseq (pl. ilisiitsit):	sorcerer
inuusuttoq (pl. inuusuttut):	a young person
isima:	thought, reason
itteq (pl. ittit):	house
iverneq (pl. ivernit):	song and drum duel
kalaalimernit:	foodstuffs which stem from the Greenlandic environment
Kalaallisut:	(West) Greenlandic language
katersutarfik:	meeting house
kiiappak:	mask, face
Kulusummeeq:	inhabitant(s) of Kulusuk
Kuummermeeq:	inhabitant(s) of Kuummiut
miartsiaq (pl. miartstsiat):	share given to somebody entitled to receive this particular share
milaarteq (pl. milaartit):	disguised player, a person in costume
milaartut:	festival at Twelfth Night (6 January)
naalanngaq:	official leader
nammeq:	up to a person; by his, her, or its own accord
nereq:	meat
piniartorssuaq:	great hunter
pisaarsiaq (pl. pisaarsiat):	present, gift
piseq (pl. pisit):	drum song

puileq (pl. puilit):	seal
pulaarteq:	visitor, guest
qattunaamiit:	store-bought products of Danish or Western origin
qattunaaq (pl. qattunat):	Danish person; sometimes used for whites or southerners more broadly
qivitteq:	half human and half non-human creature who lives in the mountains
Sermilingaarmeeq:	inhabitant(s) of Sermiligaaq
sila:	air, weather, exterior, intelligence, mind
sullivi:	service house, building offering community services
tarneq (pl. tarnit):	soul
tartaaq (pl. tartat):	shaman's helping spirit
Tasiilarmeeq:	inhabitant(s) of Tasiilaq
Tiilerilaaq:	the village of Tiniteqilaq
Tiilerilaarmeeq:	inhabitant(s) of Tiniteqilaq
timerseq:	legendary figure haunting the ice cap
timi:	body
Tunumeeq:	inhabitant(s) of East Greenland
Tunumiusut:	East Greenlandic language
tupileq (pl. tupilit):	malevolent being composed of parts of different animals, magically brought to life
uumaleq:	living being
uumasoq (pl. uumasut):	animal

Bibliography

Abrahams, R. D. 1970. A performance-centred approach to gossip. *Man* 5 (2): 290–301.

Alber, E. 2004. "Meidung als Modus des Umgangs mit Konflikten," in *Anthropologie der Konflikte. Georg Elwerts konflikttheoretische Thesen in der Diskussion*. Edited by Eckert, J., pp. 169–185. Bielefeld: Transcript Verlag.

Alia, V. 2007. *Names and Nunavut: Culture and identity in Arctic Canada*. Oxford: Berghahn.

Anderson, D. G., R. P. Wishart, and V. Vate. 2013. *About the hearth: Perspectives on the home, hearth, and household in the circumpolar north*. New York: Berghahn Books.

Aporta, C. 2004. Routes, trails and tracks: Trail breaking among the Inuit of Igloolik [Les routes, sentiers et traces chez les Inuit d'Igloolik]. *Études Inuit Studies* 28 (2): 9–38.

Apte, M. L. 1985. *Humor and laughter: An anthropological approach*. Ithaca, London: Cornell University Press.

Apte, M. L. 1987. Ethnic humor versus 'sense of humor'. *American Behavioral Scientist* 30 (3): 27–41.

Argyle, M., and J. Dean. 1965. Eye-contact, distance and affiliation. *Sociometric* 28 (3): 289–304.

Árnason, A., N. Ellison, J. Vergunst, and A. Whitehouse. Editors. 2012. *Landscapes beyond land: Routes, aesthetics, narratives*. EASA Series, Vol. 19. Oxford, New York: Berghahn.

Austin, J. L. 1962. *How to do things with words*. Oxford: Clarendon.

Bakhtin, M. M. 1981. *The dialogic imagination: Four essays*. Edited by Holquist, M. Austin: University of Texas Press.

Basso, K. H. 1970. 'To give up on words': Silence in Western Apache culture. *Southwestern Journal of Anthropology* 26: 213–230.

Basso, K. H. 1979. *Portraits of 'the Whiteman': Linguistic play and cultural symbols among the Apache*. Cambridge, London, New York, Melbourne: Cambridge University Press.

Basso, K. H. 1984. "'Stalking with stories': Names, places, and moral narratives among the Western Apache," in *Text, play, and story: The construction and reconstruction of self and society*. Edited by Plattner, S., pp. 19–55. Washington: Proceedings of the American Ethnological Society.

Bateson, G. 1972. *Steps to an ecology of mind*. New York: Balantine.

Bauman, R. 1977. *Verbal art as performance*. Rowley, MA: Newbury House Publishers, 1978.

Bergmann, J. R. 1993. *Discreet indiscretions: The social organization of gossip*. Communication and Social Order. New York: Aldine de Gruyter.

Besnier, N. 1996. "Gossip," in *Encyclopedia of cultural anthropology*. Edited by Levinson, D. and M. Ember, pp. 544–547. New York: Henry Holt & Company.

Biesele, M. 2004. "'Their own oral histories': Items of Ju/'hoan belief and items of Ju/'hoan property," in *Property and equality, vol. 1: Ritualization, sharing, egalitarianism*. Edited by Widlok, T. and W. G. Tadesse, pp. 190–200. New York, Oxford: Berghahn Books.

Binde, P. 2005. Gambling, exchange systems, and moralities. *Journal of Gambling Studies* 21 (4): 445–479.

Bird-David, N. 1987. "Single persons and social cohesion in a hunter-gatherer society," in *Dimensions of social life: Essays in honor of David G. Mandelbaum*. Edited by Mandelbaum, D. G. and P. Hockings, pp. 151–165. Berlin, New York: M. de Gruyter.

Bird-David, N. 1990. The giving environment: Another perspective on the economic system of gatherer-hunters. *Current Anthropology* 31: 189–196.

Bird-David, N. 1992. Beyond the original affluent society: A culturalist reformulation. *Current Anthropology* 33: 25–47.

Bird-David, N. 1994. Sociality and immediacy: Or, past and present conversations on bands. *Man* 29: 583–603.

Bird-David, N. 1999. 'Animism' revisited: Personhood, environment, and relational epistemology. *Current Anthropology* 40 (Supplement): S67–S91.

Birdwhistell, R. L. 1970. *Kinesics and context: Essays on body motion communication*. Philadelphia: University of Pennsylvania Press.

Bjerregaard, P., and I. Lynge. 2006. Suicide: A challenge in modern Greenland. *Archives of Suicide Research* 10 (2): 209–220.

Bodenhorn, B. 1988. Whales, souls, children, and other things that are 'good to share': Core metaphors in a contemporary whaling society. *Cambridge Anthropology* 13 (1): 1–19.

Bodenhorn, B. 1993. "Gendered spaces, public places: Public and private revisited on the North slope of Alaska," in *Landscape: Politics and perspectives*. Edited by Bender, B., pp. 169–204. Province/Oxford: Berg Publishers.

Bodenhorn, B. 1997. 'People who are like our books': Reading and teaching on the north slope of Alaska. *Arctic Anthropology* 34 (1): 117–134.

Bodenhorn, B. 2000a. "'He used to be my relative': Exploring the bases of relatedness among Iñupiat of northern Alaska," in *Cultures of relatedness: New approaches to the study of kinship*. Edited by Carsten, J., pp. 128–148. Cambridge: Cambridge University Press.

Bodenhorn, B. 2000b. "It's good to know who your relatives are but we were taught to share with everybody: Shares and sharing among Inupiaq households," in *The social economy of sharing: Resource allocation and modern hunter-gatherers*. Senri Ethnological Studies 53. Edited by Wenzel, G. W., G. K. Hovelsrud-Broda, and N. Kishigami, pp. 27–56. Osaka, Japan: National Museum of Ethnology.

Bodenhorn, B. 2004. "Is being 'Inupiat' a form of cultural property?," in *Properties of culture – culture as property: Pathways to reform in post-soviet Siberia*. Edited by Kasten, E., pp. 35–50. Berlin: Dietrich Reimer Verlag.

Bodenhorn, B. 2006. "Calling into being: Naming and speaking names on Alaska's North Slope," in *The anthropology of names and naming*. Edited by Vom Bruck, G. and B. Bodenhorn, pp. 157–176. Cambridge: Cambridge University Press.

Bordin, G. 2003. Le corpus lexical de l'habitat Inuit de l'Arctique oriental canadien. *Journal de la Société des Américanistes* 89 (1): 95–123.

Bordin, G. 2009. "Dream narration among Eastern Arctic Canadian Inuit," in *Proceedings of the 15th Inuit Studies Conference*. Edited by Beatrice, C. and M. Therrien. Paris: INALCO/CNR, www.inuitoralityconference.com/ (12.01.2010), pp. 1–15.

Bourdieu, P. 1979. "The Kabyle house or the world reversed," in *Algeria 1960: The disenchantment of the world, the sense of honour, the Kabyle house or the world reserved.* Studies in Modern Capitalism. Edited by Bourdieu, P., pp. viii, 158. Cambridge: Cambridge University Press.

Brenneis, D. L. 1984. "Straight talk and sweet talk: Political discourse in an occasionally egalitarian community," in *Dangerous words: Language and politics in the Pacific.* Edited by Brenneis, D. and F. R. Myers, pp. 69–84. Prospect Heights, IL: Waveland Press.

Brenneis, D. L. 1987. Talk and transformation. *Man* (N.S.) 22 (3): 499–510.

Briggs, J. L. 1970. *Never in anger: Portrait of an Eskimo family.* Cambridge: Harvard University Press.

Briggs, J. L. 1979. The creation of value in Canadian Inuit society. *International Journal for Social Sciences* 31 (3): 393–403.

Briggs, J. L. 1982. "Living dangerously: The contradictory foundation of value in Canadian Inuit society," in *Politics and history in band societies.* Edited by Leacock, E. B. and R. B. Lee, pp. 109–130. New York: Cambridge University Press.

Briggs, J. L. 1991. Expecting the unexpected: Canadian Inuit training for an experimental lifestyle. *Ethos* 19 (3): 259–287.

Briggs, J. L. 1993. "Lines, cycles and transformations: Temporal perspectives on Inuit action," in *Contemporary futures: Perspectives from anthropology.* ASA Monographs 30. Edited by Wellman, S., pp. 83–108. London, New York: Routledge.

Briggs, J. L. 1998. *Inuit morality play: The emotional education of a three-year-old.* Haven, CT: Yale University Press.

Briggs, J. L. 2000a. "Conflict management in a modern Inuit community," in *Hunters and gatherers in the modern world.* Edited by Schweitzer, P. P., M. Biesele, and R. K. Hitchcock, pp. 110–124. New York, Oxford: Berghahn Books.

Briggs, J. L. 2000b. Emotions have many faces: Inuit lessons. *Anthropologica* 42: 157–164.

Briggs, J. L. 2001. 'Qalluunaat run on rails; Inuit do what they want to': 'Autonomies' in camp and town. *Études Inuit Studies* 25 (1–2): 229–247.

Briggs, J. L. 2005. "Some personal thoughts on a lifelong commitment to research on Inuit culture, followed by Commentaries on the Utkuhikhalingmiutitut Dictionary," in *Building capacity in Arctic societies: dynamics and shifting perspectives.* Proceedings of the Second IPSASS Seminar (International PhD School for Studies of Arctic Societies), Iqaluit, Nunavut, CA, May 26 - June 6, 2003. Edited by Trudel, F., pp. 11-22. Québec City: CIÉRA, Faculté des sciences sociales, Université Laval.

Brison, K. J. 1992. *Just talk: Gossip, meetings, and power in a Papua New Guinea village.* Studies in Melanesian Anthropology 11. Berkeley: University of California Press.

Brown, P., and S. C. Levinson. 1978. "Universals in language use: Politeness phenomena," in *Questions and politeness.* Edited by Goody, E. N., pp. 56–289. Cambridge, London, New York, Melbourne: Cambridge University Press.

Brown, P., and S. C. Levinson. 1987. *Politeness: Some universals in language usage.* Studies in Interactional Sociolinguistics. Cambridge: Cambridge University Press.

Buijs, C. 1993. "The disappearance of traditional meat-sharing systems among the Tinitekilaamiut of East Greenland and the Arviligjuarmiut and Iglulingmiut of Canada," in *Continuity and discontinuity in Arctic cultures.* Edited by Buijs C., pp. 108–135. CNWS Publications No. 15. Leiden: National Museum of Ethnology.

Buijs, C. 2002. Familiestruktur i Østgrønland. The Greenlandic Society. www.tidsskriftet gronland.dk/archive/1987-2-Artikel03.pdf#page=14 (27.12.2005).

Buijs, C. 2004. *Furs and fabrics: Transformations, clothing and identity in East Green-land*. Leiden: CNWS Publications No. 129, Mededelingen van het Rijksmuseum voor Volkenkunde no. 32.

Buijs, C. 2010. Inuit perceptions of climate change in East Greenland. *Études Inuit Studies* 34 (1): 39–54.

Buijs, C., and J. Oosten. Editors. 1997. *Braving the cold: Continuity and change in Arctic clothing*. Leiden: CNWS Publications No. 49.

Buijs, C., and M. Petersen. 2004. Festive clothing and national costumes in 20th century East Greenland. *Études Inuit Studies* 28 (1): 83–108.

Campbell Hughes, C. 1958. Anomie, the Ammassalik, and the standardization of error. *Southwestern Journal of Anthropology* 14 (4): 352–377.

Carpenter, E. 1973. *Eskimo realities*. New York: Holt, Rinehart and Winston.

Carrithers, M., S. Collins, and S. Lukes. 1985. *The category of the person: Anthropology, philosophy, history*. Cambridge: Cambridge University Press.

Carsten, J. 2004. *After kinship*. Cambridge: Cambridge University Press.

Carsten, J., and S. Hugh-Jones. Editors. 1995. *About the house: Lévi-Strauss and beyond*. Cambridge: Cambridge University Press.

Carty, J., and Y. Musharbash. 2008. You've *got* to be joking: Asserting the analytical value of humour and laughter in contemporary anthropology. *Anthropological Forum* 18: 209–217.

Casey, E. S. 1996. "How to get from space to place in a fairly short stretch of time. Phe-nomenological prologomena," in *Senses of place*. Edited by Feld, S. and K. H. Basso, pp. 13–52. Santa Fe, New Mexico: School of American Research Press, University of Washington Press.

Celtel, A. 2005. *Categories of self: Louis Dumont's theory of the individual*. Methodology and History in Anthropology. New York, Oxford: Berghahn.

Christensen, P., and U. Bang. 2007. *Seje kvinder*. Nuuk: Milik Publishing.

Christman, J. 2004. Relational autonomy, liberal individualism, and the social constitution of selves. *Philosophical Studies* 117: 143–164.

Classen, C. 1993. *Worlds of sense: Exploring the senses in history and across cultures*. London: Routledge.

Collings, P., G. Wenzel, and R. G. Condon. 1998. Modern food sharing networks and com-munity integration in the Central Canadian Arctic. *Arctic* 51 (4): 301–314.

Corsin Jiménez, A. 2007. After trust. *Cambridge Anthropology* 25 (2): 64–78.

Critchley, S. 2002. *On humour*. London: Routledge.

Cruikshank, J. 1998. *The social life of stories: Narrative and knowledge in the Yukon ter-ritory*. Lincoln, London: University of Nebraska Press.

Cruikshank, J. 2005. *Do glaciers listen? Local knowledge, colonial encounters, and social imaginations*. Vancouver: University of British Columbia Press.

Csonka, Y. 2005. Changing Inuit historicities in West Greenland and Nunavut. *History and Anthropology* 16 (3): 321–334.

Csordas, T. J. 1994. "Introduction: The body as representation and being-in-the-world," in *Embodiment and experience, the existential ground of culture and self*. Edited by Csor-das, T. J., pp. 1–24. Cambridge: Cambridge University Press.

Csordas, T. J. 1999. "Embodiment and cultural phenomenology," in *Perspectives on embodiment: The intersections of nature and culture*. Edited by Weiss, G. and H. F. Haber, pp. 142–164. London: Routledge.

Dahl, J. 2000. *Saqqaq: An Inuit hunting community in the modern world*. Toronto: Univer-sity of Toronto Press.

Descola, P., and G. Pálsson. Editors. 1996. *Nature and society: Anthropological perspectives*. London, New York: Routledge.

Dorais, L.-J. 1981. Some notes on the language of East Greenland. *Études Inuit Studies* 5 (Supplement): 43–70.

Dorais, L.-J. 1984. Sémantique des noms d'animaux en groenlandais de l'est. *Amerindia: Revue d'Ethnolinguistique Amérindienne* 9: 7–23.

Douglas, M. 1975. *Implicit meanings: Essays in anthropology*. London: Routledge and Kegan Paul.

Douglas, M. 1999. *Implicit meanings: Selected essays in anthropology*, 2nd edition. London: Routledge.

Dowling, J. H. 1968. Individual ownership and the sharing of game in human societies. *American Anthropologist* 70: 502–507.

Downey, G. 2007. "Seeing with a 'sideways glance': Visuomotor 'knowing' and the plasticity of perception," in *Ways of knowing: New approaches in the anthropology of experience and learning*. Edited by Harris, M., pp. 222–241. Oxford: Berghahn.

Dumont, L. 1980. *Homo hierarchicus: The caste system and its implications*. 1st edition 1966. Chicago: University of Chicago Press.

Dumont, L. 1983. *Essais sur l'individualisme. Une perpectice anthropologique sur l'idéologie moderne*. Paris: Éditions du Seuil.

Duranti, A. Editor. 2009. *Linguistic anthropology: A reader*. 2nd edition. Malden, Oxford, West Sussex: Wiley-Blackwell.

Dwyer, P. D., and M. Minnegal. 2008. Fun for them, fun for us and fun for all: The 'far side' of field work in the tropical lowlands. *Anthropological Forum* 18: 303–308.

Eckert, P., and R. Newmark. 1980. Central Eskimo song duels: A contextual analysis of ritual ambiguity. *Ethnology* 19 (2): 191–211.

Eisenberg, A. R. 1986. "Teasing: Verbal play in two Mexican homes," in *Language socialization across cultures*. Studies in the Social and Cultural Foundation of Language No. 3. Edited by Schieffelin, B. B. and E. Ochs, pp. 182–198. Cambridge, New York, Melbourne: Cambridge University Press.

Elixhauser, S. 2006. *Ethik in der angewandten Ethnologie. Eine Feldforschung zum Tourismus auf den Philippinen*. Fokus Kultur – Trierer Beiträge zur gegenwartsbezogenen Ethnologie. Norderstedt: BoD.

Elixhauser, S. 2009a. "Filmen im Familienalltag. Sermiligaaq 65° 54' N, 36° 22' W – ein ethnografisches Dokumentarfilmprojekt in Ostgrönland (2006–2008)," in *Die arktische Leinwand. Grönland im Film. Von Kund Rasmussen bis "Fräulein Smilla". 51. Nordische Filmtage Lübeck 2009*. Edited by Hamburg, H. L. K., pp. 80–98. Lübeck: Schmidt-Römhild.

Elixhauser, S. 2009b. Krokodile in Grönland. Humor im interkulturellen Vergleich: ein Forschungsthema in der Ethnologie. *Mondial – SIETAR Journal für interkulturelle Perspektiven* 2: 3–7.

Elixhauser, S. 2015. Travelling the East Greenlandic sea- and landscape: Encounters, places, and stories. *Mobilities* 10 (4): 531–551.

Elixhauser, S. in press-a. "Inuit responses to arctic militarization: Examples from East Greenland," in *Exploring ice and snow in the Cold War: Histories of extreme climatic environments*. Edited by Herzberg, J., C. Kehrt, and F. Torma. New York, Oxford: Berghahn.

Elixhauser, S. in press-b. "Moving along: Wayfinding and non-verbal communication across the frozen seascape of East Greenland," in *At home on the waves: Human habitation of the sea from the Mesolithic to today*. Edited by King, T. J. and G. Robinson. New York, Oxford: Berghahn.

Elixhauser, S. forthcoming. "Unterwegs im Eismeer: Mobilität, persönliche Autonomie und sozialer Zusammenhalt im ländlichen Ostgrönland," in *Grönland. Kontinuitäten und Brüche im Leben der grönländischen Inuit.* Edited by Sowa, F. Leverkusen: Budrich UniPress.

Emdal Navne, L. 2008. Flexibelt moderskab. Reproduktive beslutninger, livsforløb og slægtskab i Grønland. M.A. Thesis (Kandidatspeciale), Copenhagen: University of Copenhagen.

Endicott, K. 1998. "Property, power and conflict among the Batek of Malaysia," in *Hunters and gatherers: Property, power and ideology.* Vol. 2. Edited by Ingold, T., D. Riches, and J. Woodburn, pp. 110–127. Oxford, Washington, DC: Berg Publishers.

Endicott, K. L. 1999. "Gender relations in hunter-gatherer societies," in *The Cambridge encyclopedia of hunters and gatherers.* Edited by Lee, R. and R. H. Daly, pp. 411–418. Cambridge: Cambridge University Press.

Enel, C., and S. Basbøll. 1981. Contribution à la découverte d'Ammassalik: Adolf Erik Nordenskiöd précède Gustav Holm. *Études Inuit Studies* 5 (1): 126–129.

Farnell, B. 1995. *Do you see what I mean? Plain Indian sign talk and the embodiment of action.* Austin: University of Texas Press.

Farnell, B. 1999. Moving bodies, acting selves. *Annual Review of Anthropology* 28: 341–373.

Farnell, B. 2003. Kinesthetic sense and dynamically embodied action. *Journal for the Anthropological Study of Human Movement* 12 (4). www.findarticles.com/p/articles/mi_qa4093/is_200310/ai_n9284163/print (27.12.2005).

Feit, H. A. 1996. "The enduring pursuit: Land, time, and social relationships in anthropological models of hunter-gatherers and subarctic hunters' images," in *Key issues in hunter gatherer research.* Edited by Burch, E. S. J. and L. J. Ellana, pp. 421–439. Oxford, Washington: Berg Publishers.

Ferraro, G. P. 2006. *Cultural anthropology: An applied perspective.* 6th edition. Belmont, CA: Thomson/Wadsworth.

Fienup-Riordan, A. 1986. The real people: The concept of personhood among the Yup'ik Eskimos of Western Alaska. *Études Inuit Studies* 10 (1–2): 261–270.

Fienup-Riordan, A. 1994. *Boundaries and passages: Rule and ritual in Yup'ik Eskimo oral tradition.* The Civilization of the American Indian Series, Vol. 212. Norman: University of Oklahoma Press.

Fienup-Riordan, A. 2005. *Wise words of the Yup'ik people: We talk to you because we love you.* Lincoln, NE: University of Nebraska Press, Chesham.

Fikentscher, W. 2008. "Chapter 04: Social norms," in *Law and anthropology: Outlines, issues, suggestions,* pp. 1–12. Munich: Bayerische Akademie der Wissenschaften, C. H. Beck. http://works.bepress.com/wolfgang_fikentscher/4 (10.03.2010).

Finnegan, R. H. 2002. *Communicating: The multiple modes of human interconnection.* London, New York: Routledge.

Flora, J. 2012. "'I don't know why he did it: It happened by itself': Causality and suicide in Northwest Greenland," in *The anthropology of ignorance: An ethnographic approach.* Edited by High, C., A. Kelly, and J. Mair, pp. 137–161. New York, NY: Palgrave Macmillan.

Flora, J. 2015. "The lonely un-dead and returning suicide in Northwest Greenland," in *Suicide and agency: Anthropological perspectives on self-destruction, personhood, and power.* Edited by Münster, D. and L. Brozek, pp. 47–66. London, New York: Routledge.

Fortescue, M. 1988. Eskimo orientation system. *Meddelelser om Grønland, Man & Society* 11: 3–30.

Fowler, C. S., and N. J. Turner. 1999. "Ecological/cosmological knowledge and land management among hunter-gatherers," in *The Cambridge encyclopedia of hunters and gatherers*. Edited by Lee, R. B. and R. Daly, pp. 419–425. Cambridge: Cambridge University Press.

Freeman, M. R. 1996. "Why mattak and other kalaalimerngit [local foods] matter," in *Cultural and social research in Greenland 95/96: Essays in honour of Robert Petersen*. Edited by Jacobsen, B., C. Andreasen, and J. Rygaard, pp. 45–53. Nuuk: Ilisimatusarfik/Atuakkiorfik.

Freud, S. 1960. *Jokes and their relation to the unconscious*. New York: W. W. Norton (original work published 1905).

Gambetta, D. Editor. 1988. *Trust: Making and breaking co-operative relations*. Oxford: Blackwell.

Gardner, P. M. 1991. Foragers' pursuit of individual autonomy. *Current Anthropology* 32: 543–572.

Geertsen, I. 1994. *Grønlandske masker = Greenlandic masks*. Copenhagen: Rhodos.

Gessain, R. 1967. Angmassalik, trente ans après. Evolution d'une tribu eskimo dans le monde moderne. *Objets et Mondes* 7 (2): 133–156.

Gessain, R. 1968. The Ammassalimiut kayak and its demographic evolution: Extract from *Objets et Mondes* 8 (part four). www.arctickayaks.com/PDF/Gessain1968/gessain-pt1.htm (01.07.2005).

Gessain, R. 1969. *Ammasalik où la civilisation obligatoire*. Paris: Flammarion.

Gessain, R. 1970. *Angmagssalik. Den påtvungene civilisation*. Copenhagen: Rhodos.

Gessain, R. 1978. "L'Homme-Lune dans la mythologie des Ammmassalimiut," in *Systèmes de signes: textes réunis en hommage à Germaine Dieterlen*. Edited by Dieterlen, G., pp. 205–222. Paris: Herman.

Gessain, R. 1980. Nom et réincarnation chez les Ammassalimiut. *Boréales, Revue du Centre de Recherches Inter-Nordique* 15–16: 407–419.

Gessain, R. 1984. Dance masks of Ammassalik (East coast of Greenland). *Arctic Anthropology* 21 (2): 81–107.

Gessain, R., L.-J. Dorais, and C. Enel. 1986. *Vocabulaire du groenlandais de l'est. 1440 mots de la langue des Ammassalimiut avec leur traduction en groenlandais de l'ouest, francais, anglais et danois*. Documents du Centre de Recherches Anthropologiques du Musée de l'Homme. No. 10. Paris.

Gessain, R., and P.-É. Victor. 1969. Le kayak des Ammassalimiut. *Extrait de la revue 'Objets et Mondes'* 9 (2): .145–160.

Gessain, R., and P.-É. Victor. 1973. Le tambour chez les Ammassalimiut (côte est du Groënland). *Extrait de la revue 'Objets et Mondes'* 13 (3): 129–160.

Gessain, R., and P.-É. Victor. 1974. Magitärneq jeu d'osselets chez les Ammassalimiut (côte est du Groënland). *Extrait de la revue 'Objets et Mondes'* 14 (2): 73–88.

Gibson, J. J. 1979. *The ecological approach to visual perception*. Boston: Houghton Mifflin.

Gluckman, M. 1963. Gossip and scandal. *Current Anthropology* 4: 307–315.

Gluckman, M. 1968. Psychological, sociological and anthropological explanations of witchcraft and gossip: A clarification. *Man* 3 (1): 20–34.

Goehring, B., and J. K. Stager. 1991. The intrusion of industrial time and space into the Inuit lifeworld: Changing perceptions and behaviour. *Environment and Behavior* 23 (6): 666–679.

Goffman, E. 1955. On face-work: An analysis of ritual elements of social interactions. *Psychiatry: Journal for the Study of Interpersonal Processes* 18 (3): 213–231.

Goffman, E. 1959. *The presentation of self in everyday life*. New York: Doubleday & Company.

Goffman, E. 1967. *Interaction ritual: Essays in face-to-face behavior*. Chicago: Aldine Pub. Co.

Gombay, N. 2009. "'Today is today and tomorrow is tomorrow': Reflections on Inuit understanding of time and place," in *Proceedings of the 15th Inuit Studies Conference*. Edited by Beatrice, C. and M. Therrien. Paris: INALCO/CNR. www.inuitoralityconference. com/ (12.01.2009), pp. 1–15.

Goodale, M. A., and A. D. Milner. 2004. *Sight unseen: An exploration of conscious and unconscious vision*. Oxford: Oxford University Press.

Goodale, M. A., and D. A. Westwood. 2004. An evolving view of duplex vision: Separate but interacting cortical pathways for perception and action. *Current Opinion in Neurobiology* 14: 203–211.

Goodman, J. E., and P. A. Silverstein. 2009. *Bourdieu in Algeria: Colonial politics, ethnographic practices, theoretical developments*. Lincoln, NE: University of Nebraska Press; Chesham: Combined Academic [distributor].

Goodwin, C. 1990. Conversation analysis. *Annual Review of Anthropology* 18: 283–307.

Goody, E. N. 1978. "Towards a theory of questions," in *Questions and politeness*. Edited by Goody, E. N., pp. 17–43. Cambridge, London, New York, Melbourne: Cambridge University Press.

Gosden, C., and C. Knowles. 2001. "People, objects and colonial relations," in *Collecting colonialism: Material culture and colonial change*, pp. 1–25. Oxford: Berg.

Grasseni, C. 2004. Skilled vision: An apprenticeship in breeding aesthetics. *Social Anthropology* 12 (1): 41–55.

Grasseni, C. 2007a. "Communities of practice and forms of life: Towards a rehabilitation of vision?," in *Ways of knowing: Anthropological approaches to crafting experience and knowledge*. Methodology and History in Anthropology. Edited by Harris, M., pp. 203–221. New York, Oxford: Berghahn Books.

Grasseni, C. Editor. 2007b. *Skilled visions: Between apprenticeship and standards*. EASA Series. Oxford: Berghahn.

Griffin, D. R. 2001. *Animal minds: Beyond cognition to consciousness*. Chicago, London: University of Chicago Press.

Guemple, D. L. 1979. *Inuit adoption*. Ottawa: National Museums of Canada.

Guemple, D. L. 1993. "Born-again pagans: The Inuit cycle of spirits," in *Amerindian rebirth: Reincarnation belief among North American Indians and Inuit*. Edited by Mills, A. C. and R. Slobodin, pp. 107–122. Toronto: University of Toronto Press.

Guemple, D. L. 1995. "Gender in Inuit society," in *Women and power in native North America*. Edited by Ackerman, L. A. and L. F. Klein, pp. 17–27. Norman, London: University of Oklahoma Press.

Gumperz, J. J., and D. Hymes. 1986. *Directions in sociolinguistics: The ethnography of communication*. Oxford: Basil Blackwell.

Hall, E. T. 1959. *The silent language*. New York: Doubleday.

Hall, E. T. 2003. "Proxemics," in *The anthropology of space and place: Locating culture*. Edited by Low, S. M. and D. Lawrence-Zuniga, pp. 51–73. Oxford: Blackwell Publishing.

Hallowell, A. I. 1955. *Culture and experience*. Publications of the Philadelphia Anthropological Society, Vol. 4. Philadelphia: University of Pennsylvania Press.

Hallowell, A. I. 1960. "Ojibwa ontology, behavior and world view," in *Culture in history: Essays in honor of Paul Radin*. Edited by Dimond, S., pp. 19–52. New York: Columbia University Press.

Handelman, D. 1973. Gossip in encounters: The transmission of information in a bounded social setting. *Man* 8 (2): 210–227.

Hann, C. M. 1998a. "Introduction: The embeddedness of property," in *Property relations: Renewing the anthropological tradition*. Edited by Hann, C. M., pp. 1–47. Cambridge: Cambridge University Press.

Hann, C. M. Editor. 1998b. *Property relations: Renewing the anthropological tradition*. Cambridge: Cambridge University Press.

Haraway, D. 1988. Situated knowledge: The science question in feminism and the privilege of partial perspective. *Feminist Perspective* 14 (3): 575–599.

Hargie, O., and D. Dickson. 2004. *Skilled interpersonal communication: Research, theory and practice*. New York: Routledge.

Harré, R. 1984. *Personal being: A theory for individual psychology*. Cambridge, MA: Harvard University Press.

Harré, R. 1998. *The singular self: An introduction to the psychology of personhood*. London, Thousand Oaks, New Delhi: Sage Publications.

Harvey, G. 2005. *Animism: Respecting the living world*. London: Hurst & Co.

Haviland, J. B. 1977. *Gossip, reputation, and knowledge in Zinacantan*. Chicago, London: University of California Press.

Haviland, J. B., and L. K. Haviland. 1983. "Privacy in a Mexican Indian village," in *Public and private in social life*. Edited by Benn, S. I. and G. F. Gaus, pp. 341–362. London: Croom Helm.

Hawkes, E. W. 1916. *The Labrador Eskimo*. Ottawa: Government Printing Bureau.

Helliwell, C. 1995. Autonomy as natural equality: Inequality in 'egalitarian' societies. *The Journal of the Royal Anthropological Institute* 1: 359–375.

Hendry, J. 1989. To wrap or not to wrap: Politeness and penetration in ethnographic inquiry. *Man* (N.S.) 24 (4): 620–635.

Hendry, J., and C. W. Watson. Editors. 2001. *The anthropology of indirect communication*. ASA Monographs 37. London, New York: Routledge.

Henriksen, G. 1993. *Hunters in the Barrens: The Naskapi on the edge of the White Man's world*. Newfoundland Social and Economic Studies No. 12. Institute of Social and Economic Research, St. Johns: Memorial University of Newfoundland.

Hess, S. 2006. Strathern's Melanesian 'dividual' and the Christian 'individual': A perspective from Vanua Lava, Vanuatu. *Oceania* 76 (3): 285–296.

Hicks, J. 2007. "Suicide by Greenlandic youth, in historical and circumpolar perspective," in *Children and youth in Greenland: An anthology*. Edited by MIPI. Nuuk: Ilisimatusarfik and MILIK Publishing. https://alaskaindigenous.files.wordpress.com/2012/07/hicks-sucide-in-greenland.pdf (24.06.2017).

Hindsberger, M. 1999. *Fra åndetro til gustro i Østgrønland*. København: Museum Tusculanums Forlag, Københavns Universitet.

Hirsch, E., and M. O'Hanlon. Editors. 1995. *The anthropology of landscape: Perspectives on place and space*. Oxford: Clarendon Press.

Hoebel, E. A. 1966. *Anthropology: The study of man*. 3rd edition. New York: McGraw-Hill.

Holm, G. 1888. Ethnografisk skizze af Ammassalikerne. Den Ostgrønlandske expedition, udført I aarene 1883–1885. *Meddelelser om Grønland* 10: 43–182.

Holm, G. 1911. "Ethnological sketch of the Angmagsalik Eskimo," in *The Ammassalik Eskimo: Contributions to the ethnology of the East Greenland natives*. Vol. 39 (1), 1914. Edited by Thalbitzer, W., pp. 1–148. Transl. from Danish version 1888. Copenhagen: Meddelelser om Grønland.

Holm, G., and J. Petersen. 1912 (1887). Legends and tales from Angmagsalik collected by G. Holm and translated by Johan Petersen. *Meddelelser om Grønland* 39 (4): 225–306.

Hornborg, A., and G. Pálsson. 1996. Nature and society: A contested interface. *Lundalinjer* 114: 43–57.

Hovelsrud-Broda, G. K. 1997. Arctic seal-hunting households and the anti-sealing controversy. *Research in Economic Anthropology* 18: 17–34.

Hovelsrud-Broda, G. K. 1999a. "Contemporary seal hunting households: Trade bans and subsidies," in *At the interface: The household and beyond*. Monography in Economic Anthropology No. 15. Edited by Small, D. B. and N. Tannenbaum, pp. 95–112. Lanham, New York, Oxford: University Press of America, Inc.

Hovelsrud-Broda, G. K. 1999b. The integrative role of seals in an East Greenlandic hunting village. *Arctic Anthropology* 36 (1–2): 37–50.

Hovelsrud-Broda, G. K. 2000a. "The isertormeeq of East Greenland," in *Endangered peoples of the Arctic: Struggles to survive and thrive*. Edited by Freeman, M. M. R., pp. 151–166. Westport, CT: Greenwood Press.

Hovelsrud-Broda, G. K. 2000b. "'Sharing', transfers, transactions and the concept of generalized reciprocity," in *The social economy of sharing: Resource allocation and modern hunter-gatherers*. Senri Ethnological Studies 53. Edited by Wenzel, G. W., G. K. Hovelsrud-Broda, and N. Kishigami, pp. 193–214. Osaka, Japan: National Museum of Ethnology.

Howe, J., and J. Sherzer. 1986. Friend hairyfish and friend rattlesnake, or, keeping anthropologists in their place. *Man* (N.S.) 21: 680–696.

Humphrey, C. 1988. No place like home in anthropology: The neglect of architecture. *Anthropology Today* 4 (1): 16–18.

Hymes, D. H. 1962. "The ethnography of speaking," in *Anthropology and human behavior*. Edited by Gladwin, T. and W. C. Sturtevant, pp. 13–53. Washington: Anthropological Society of Washington.

Hymes, D. H. 1964. Introduction: Towards ethnographies of communication. *American Anthropologist* 66 (6, part 2): 1–34.

Hymes, D. H., and J. J. Gumperz. Editors. 1964. The ethnography of communication (special issue). *American Anthropologist* 66 (6, part 2): 1–186.

Ingold, T. 1986. *The appropriation of nature: Essays on human ecology and social relations*. Iowa City: University of Iowa Press.

Ingold, T. 1991. Becoming persons: Consciousness and sociality in human evolution. *Cultural Dynamics* 4: 355–378.

Ingold, T. 1993a. "The art of translation in a continuous world," in *Beyond boundaries: Understanding, translation, and anthropological discourse*. Explorations in Anthropology. Edited by Pálsson, G., pp. 210–230. Oxford: Berg.

Ingold, T. 1993b. The temporality of landscape. *World Archaeology* 25 (2): 152–174.

Ingold, T. 1994. "Introduction," in *What is an animal?* Edited by Ingold, T., pp. 1–15. Cambridge: Cambridge University Press.

Ingold, T. 1997. "Life beyond the edge of nature? Or, the mirage of society," in *The mark of the social*. Edited by Greenwood, J. D., pp. 231–252. Lanham, MD: Rowman & Littlefield Publishers.

Ingold, T. 1999a. Human nature and science. *Interdisciplinary Science Reviews* 24 (4): 250–254.

Ingold, T. 1999b. "On the social relations of the hunter-gatherer band," in *The Cambridge encyclopedia of hunters and gatherers*. Edited by Lee, R. and R. H. Daly, pp. 399–409. Cambridge: Cambridge University Press.

Ingold, T. 2000. *The perception of the environment: Essays on livelihood, dwelling and skill*. London, New York: Routledge.

Ingold, T. 2004a. Culture on the ground: The world perceived through the feet. *Journal of Material Culture* 9 (3): 315–340.

Ingold, T. 2004b. "Time, memory and property," in *Property and equality, vol. 1: Ritualization, sharing, egalitarianism*. Edited by Widlok, T. and W. G. Tadesse, pp. 165–174. New York, Oxford: Berghahn Books.

Ingold, T. 2006. Rethinking the animate, re-animating thought. *Ethnos* 71 (1): 9–20.

Ingold, T. 2007. *Lines: A brief history*. London: Routledge.

Ingold, T. 2011. *Being alive: Essays on movement, knowledge and description*. London: Routledge.

Ingold, T., and J. L. Vergunst. Editors. 2008. *Ways of walking: Ethnography and practice on foot*. Anthropological Studies of Creativity and Perception. Aldershot: Ashgate.

Inness, J. C. 1992. *Privacy, intimacy, and isolation*. New York, Oxford: Oxford University Press.

Jackson, M. 1996. *Things as they are: New directions in phenomenological anthropology*. Bloomington: Indiana University Press.

Jacobsen, B. 2006. "Grønlandsk chat – sprogmøde i cyberspace," in *Grønlandsk kultur- og samfundsforskning 2006–07*. Edited by Ilisimatusarfik, pp. 115–126. Nuuk: Atuagkat.

Jakobsen, M. D. 1999. *Shamanism: Traditional and contemporary approaches to the mastery of spirits and healing*. New York, Oxford: Bergham Books.

Jaworski, A. 1993. *The power of silence: Social and pragmatic perspectives*. Language and Language Behaviors. Newbury Park, CA, London: Sage.

Jaworski, A. Editor. 1997. *Silence: Interdisciplinary perspective*. Studies in Anthropological Linguistics. Berlin, New York: Mouton de Gruyter.

Kellenberger, J. 1984. Kierkegaard, indirect communication, and religious truth. *International Journal for Philosophy of Religion* 16: 153–116.

Kielsen, L. 1996. *Mitaartut*: An Inuit winter festival in Greenland. *Études Inuit Studies* 20 (1): 123–129.

Kierkegaard, S. 1992. *Concluding unscientific postscript to philosophical fragments*. Vol. 1. Princeton, NJ: Princeton University Press.

Kingston, D. P. 2008. The persistence of conflict avoidance among the King Island Inupiat. *Études Inuit Studies* 32 (2): 151–167.

Kleivan, I. 1960. Mitârtut: Vestiges of the Eskimo sea-woman cult in West Greenland. *Meddelelser om Grønland* 161 (5): 1–30.

Kleivan, I. 1971. Song duels in West Greenland – joking relationship and avoidance. *Folk* 13: 9–13.

Kleivan, I., and B. Sonne. 1985. *Eskimos Greenland and Canada*. Iconography of Religions, Section VIII: Arctic Peoples. Leiden: Institute of Religious Iconography, State University Groningen, E.J. Brill.

Kommuneqarfik Sermersooq. 2016. "Lokalsamfundsprofil Sermiligaaq," 2. udgave 2016. https://sermersooq.gl/uploads/2016/10/FINAL-Sermiligaaq-Lokalsamfundsprofil-DK-2016.pdf (24.05.2017).

Korta, K. 2008. Malinowski and pragmatics: Claim making in the history of linguistics. *Journal of Pragmatics* 40 (10): 1645–1660.

Krogh Andersen, M. 2008. *Grønland. Mægtig of afmægtig*. Copenhagen: Gyldendal, Nordisk Forlag.

Kulick, D., and C. Strout. 1990. Christianity, cargo and ideas of self: Patterns of literacy in a Papua New Guinean village. *Man* 25 (2): 286–304.

Kusenbach, M. 2003. Street phenomenology: The go-along as ethnographic research tool. *Ethnography* 4 (3): 455–485.

Kwon, H. 1998. The saddle and the sledge: Hunting as comparative narrative in Siberia and beyond. *The Journal of the Royal Anthropological Institute* 4: 115–127.

Lakoff, R. T. 2007. "The triangle of linguistic structure," in *A cultural approach to interpersonal communication: Essential readings*. Edited by Monaghan, L. and J. E. Goodman, pp. 128–133. Malden, Oxford, Carlton: Blackwell Publishing.

Landes, J. B. 1998. *Feminism, the public and the private*. New York: Oxford University Press.

Langgård, K. 1998. "Vestgrønlændernes syn på østgrønlænderne gennem tiden," in *Grønlandsk kultur- og Samfundsforskning 98/99*. Edited by Ilisimatusarfik, pp. 175–200. Nuuk: Atuagkat.

Larsen, F. B. 1992a. "Death and social change in Ittoqqortoormiit (East Greenland)," in *Regard sur l'avenir/Looking to the future. Communications du 7e Congrès d'Études Inuit/ Papers from the 7th Inuit studies Conference*. Edited by Dufour, M.-J. and F. Thérien, pp. 125–139. Québec: Université Laval.

Larsen, F. B. 1992b. "Voldsom død og social forandring Iittoqqortoormiit," in *Grønlandsk Kultur- og Samfundsforskning 92*. Edited by Ilisimatusarfik, pp. 129–146. Nuuk: Atuagkat.

Latour, B. 1993. *We have never been modern*. New York: Harvester Wheatsheaf.

Laugrand, F., and J. G. Oosten. 2008. *The sea woman: Sedna in Inuit shamanism and art in the eastern Arctic*. Fairbanks: University of Alaska Press.

Laugrand, F., J. G. Oosten, and W. Rasing. 1999a. "Introduction: *Tirigusuusiit, piqujait* and *maligait*: Inuit perspectives on traditional law," in *Perspectives on traditional law*. Interviewing Inuit Elders Series. Edited by Laugrand, F., J. Oosten, and W. Rasing, pp. 1–12. Iqaluit: Language and Culture Program, Nunavut Arctic College.

Laugrand, F., J. G. Oosten, and W. Rasing. Editors. 1999b. *Perspectives on traditional law*. Interviewing Inuit Elders Series. Iqaluit: Language and Culture Program, Nunavut Arctic College.

Laugrand, F., J. G. Oosten, and F. Trudel. 2002. Hunters, owners, and givers of light: The tuurngait of South Baffin Island. *Arctic Anthropology* 39 (1–2): 27–50.

Lave, J. 1988. *Cognition in practice: Mind, mathematics and culture in everyday life*. Cambridge: Cambridge University Press.

Law, S., and L. J. Kirmayer. 2005. Inuit interpretations of sleep paralysis. *Transcultural Psychiatry* 42: 93–112.

Lee, J., and T. Ingold. 2006. "Fieldwork on foot: Perceiving, routing, socializing," in *Locating the field: Space, place and context in anthropology*. Edited by Coleman, S. and P. Collins, pp. 67–85. Oxford: Berg.

Lee, M., G. A. Reinhardt, and A. Tooyak. 2003. *Eskimo architecture: Dwelling and structure in the early historic period*. Fairbanks: University of Alaska Press and University of Alaska Museum.

Lee, R. B., and I. DeVore. Editors. 1968. *Man the hunter*. Chicago: Aldine Pub. Co.

Legat, A. 2012. *Walking the land, feeding the fire. Knowledge and stewardship among the Tlicho Dene*. Tucson: University of Arizona Press.

Leineweber, M. J. 2000. Modernization and mental health: Suicide among the Inuit in Greenland. M.A. Thesis, Nijmegen: Katholieke Universiteit Nijmegen.

Lévi-Strauss, C. 1987. *Lévi-Strauss, anthropology and myth: Lectures 1951–1982*. Oxford: Blackwell.

Lynge, F. 1992. *Arctic wars, animal rights, endangered peoples*. Translated by Marianne Stenbaek. Hannover, London: Darthmouth College Press.

MacDonald, C. J.-H. 2008. Cooperation, sharing and reciprocity (sharing without giving, receiving without owing). https://sites.google.com/site/charlesjhmacdonaldssite/ (12.01.2010), pp. 1–21.

MacDonald, C. J.-H. 2009. "Inuit personal names: A unique system?" in *Proceedings of the 15th Inuit Studies Conference*. Edited by Beatrice, C. and M. Therrien. Paris: INALCO/ CNR. www.inuitoralityconference.com/ (12.01.2009), pp. 1–17.

MacDonald, J. 1998. *The Arctic sky: Inuit astronomy, star lore, and legend.* Toronto: Royal Ontario Museum and Nunavut Research Institute.

Machin, B., and W. Lancaster. 1974. Correspondence: Anthropology and gossip. *Man* 9 (4): 625–627.

Madden, K. 1997. Asserting a distinctly Inuit concept of news: The early years of the Inuit broadcasting corporation's Qagik. *Howard Journal of Communications* 8 (4): 297–313.

Malinowski, B. 1923. "The problem of meaning in primitive language: Supplement," in *The meaning of meaning: A study of the influence of language upon thought and of the science of symbolism.* 10th edition 1972. Edited by Ogden, C. K. and I. A. Richard, pp. 296–336. London: Routledge and Kegan Paul.

Malinowski, B. 1935. *Coral gardens and their magic: A study of the methods of tilling the soil and of agricultural rites in the Trobriand Islands, vol. 1: The description of gardening; vol. 2: The language of magic and gardening.* London: Allen & Unwin.

Maqe, E., and C. Enel. 1993a. *Elisa Maqe-p oqaluttuarai ammit pissamaatit/Elisa Maqe fortæller om skind of forråd.* Translated and written down by Catherine Enel. 2nd edition 2002. Tasiilaq: Tassilap Katersugaasivia/Ammassalik Museum.

Maqe, E., and C. Enel. 1993b. *Elisa Maqe-p oqaluttuarai illuttoqqt tupitoqqart/Elisa Maqe fortæller om det traditionelle hus det traditionelle tent.* Translated and written down by Catherine Enel. Tasiilaq: Tassilap Katersugaasivia/Ammassalik Museum.

Maqe, E., and J. Rosing. 1994. *Tunumiit mersertini oqalittuaat/Østgrønlandske børneeventyr/East Greenlandic children's stories.* Nuuk: Atuakkiorfik.

Mather, E. P., and P. Morrow. 1998. There are no more words to the story. *Oral Tradition* 13 (1): 300–342.

Mauss, M. 1906. *Essai sur les variations saisonnières des sociétés eskimos. Études de morphologie sociale.* L'Anné sociologique 1904–1905, with the collaboration of Beuchat, H. IX. Paris: Alcan.

Mauss, M. 1923–1924. Essai sur le don. Forme et raison de l'échange dans les sociétés primitives. Originally published in: *L'Année Sociologique, seconde série, 1923-1924.* http:// nthropomada.com/bibliotheque/Marcel-MAUSS-Essai-sur-le-don.pdf (10.10.2017).

Mauss, M. 1985. "A category of the human mind: The notion of person; the notion of self (Translated by Halls, W. D.)," in *The category of the person: Anthropology, philosophy, history.* Edited by Carrithers, M., S. Collins, and S. Lukes, pp. 1–25. Cambridge: Cambridge University Press.

Mazzullo, N., and T. Ingold. 2008. "Being along: Place, time and movement among Sámi people," in *Mobility and place: Enacting Northern European peripheries.* Edited by Bærenholdt, J. O. and B. Granas, pp. 27–38. Aldershot: Ashgate.

Mead, G. H. 1934. *Mind, self, and society from the standpoint of a social behaviorist.* Edited by Morris, C. W. Chicago: University of Chicago Press.

Mejer, S. 2007. "Socialisation i Østgrønland i nyere tid," in *Børn og unge i Grønland – en antologi.* Edited by Kahlig, W. and N. Banerjee, pp. 176–193. Nuuk: MIPI Videnscenter om Børn og Unge, Ilisimatusarfik University of Greenland, Milik Publishing.

Mennecier, P. 1995. *Le tunumiisut, dialecte inuit du Groenland oriental. Description et analyse*. Paris: Société de Linguistique de Paris.

Merleau-Ponty, M. 1962. *Phenomenology of perception*. Translated by Colin Smith. London: Routledge & Kegan Paul.

Merleau-Ponty, M. 1964. "Eye and mind," in *The primacy of perception, and other essays on phenomenological psychology, the philosophy of art, history and politics*. Edited by Evanston, E. J. M., pp. 159–190. Evanston: Northwestern University Press.

Mikkelsen, E. 1925. *Med 'Grønland' til Scoresby Sund*. Copenhagen: Gyldendal.

Mikkelsen, E., in collaboration with P. P. Sveistrup. 1944. *The East Greenlanders possibilities of existence, their production and consumption*. Meddelser om Gronland Bd. 134 Nr. 2. Copenhagen: C.A. Reitzels Forlag.

Milton, K. 2002. *Loving nature: Towards an ecology of emotion*. London: Routledge.

Mitchell, W. E. 1988. The defeat of hierarchy: Gambling as exchange in a Sepik society. *American Ethnologist* 15 (4): 638–657.

Mollerup, P. 2006. *Wayshowing: A guide to environmental signage, principles & practices*. Baden: Lars Muller.

Moore, A. D. 2003. Privacy: Its meaning and value. *American Philosophical Quarterly* 40 (3): 215–227.

Morreal, J. 1994. "Gossip and humour," in *Good gossip*. Edited by Goodman, R. F. and A. Ben-Ze'ev, pp. 56–64. Lawrence, KS: University Press of Kansas.

Morris, B. 1994. *Anthropology of the self: The individual in cultural perspective*. London, Boulder, CO: Pluto Press.

Morris, M. 1981. *Saying and meaning in Puerto Rico: Some problems in the ethnography of discourse*. Oxford: Pergamon Press.

Morrow, P. 1990. Symbolic actions, indirect expressions: Limits to interpretations of Yupik society. *Études Inuit Studies* 14 (1–2): 141–158.

Morrow, P. 1995. "On shaky grounds: Folklore, collaboration, and problematic outcomes," in *When our words return*. Edited by Morrow, P. and W. Schneider, pp. 27–51. Logan: Utah State University Press.

Morrow, P. 1996. Yup'ik Eskimo agents and American legal agencies: Perspectives on compliance and resistance. *Journal of the Royal Anthropological Institute* 2 (3): 405–423.

Myers, F. R. 1986. *Pintupi country, Pintupi self: Sentiment, place, and politics among Western Desert Aborigines*. Washington, DC: Smithsonian Institution Press.

Myers, F. R. 1988. Critical trends in the study of hunter-gatherers. *Annual Review of Anthropology* 17: 261–281.

Nellemann, G. 1960. Mitârneq: A West Greenland winter ceremony. *Folk* 2: 99–113.

Nelson, R. K. 1969. *Hunters of the northern ice*. Chicago: University of Chicago Press.

Nelson, R. K. 1983. *Make prayers to the raven: A Kuyokon view of northern forest*. Chicago: University of Chicago Press.

Nooter, G. 1972/73. Change in a hunting community in East Greenland. *Folk* 14–15: 163–204.

Nooter, G. 1975. Mitârtut, winter feast in Greenland. *Objets et Mondes* 15 (2): 159–169.

Nooter, G. 1976. *Leadership and headship: Changing authority patterns in an East Greenland hunting community*. Leiden: E. J. Brill.

Nooter, G. 1984a. "1884–1984: A century of changes in East Greenland," in *Life and survival in the Arctic: Cultural changes in the Polar regions*. Ethnological serie Verre naasten naderbij. Edited by Nooter, G., pp. 121–143. Den Haag: Staatsuitgeverij, 's-Gravenhage.

Nooter, G. Editor. 1984b. *Life and survival in the Arctic: Cultural changes in the Polar regions.* Ethnological serie Verre naasten naderbij. Den Haag: Staatsuitgeverij, 's-Gravenhage.

Nooter, G. Editor. 1984c. *Over leven en overleven. Culturveranderingen in de poolgebieden.* Volkenkundige reeks Verre nasten naderbij. Den Haag: Staatsuitgeverij, 's Gravenhage.

Nooter, G. 1988. Some recent developments in the Ammassalik district, East Greenland. *Folk* 30: 215–228.

Nuttall, M. 1991. Memoryscape: A sense of locality in Northwest Greenland. *North Atlantic Studies* 1 (2): 39–50.

Nuttall, M. 1992. *Arctic homeland: Kinship, community and development in Northwest Greenland.* Toronto, Buffalo: University of Toronto Press.

Nuttall, M. 1993. "The name never dies: Greenland Inuit ideas of the person," in *Amerindian rebirth: Reincarnation belief among North American Indians and Inuit.* Edited by Mills, A. C. and R. Slobodin, pp. 123–135. Toronto: University of Toronto Press.

Nuttall, M. 2000a. Becoming a hunter in Greenland. *Études Inuit Studies* 24 (2): 33–47.

Nuttall, M. 2000b. "Choosing kin: Sharing and subsistence in a Greenlandic hunting community," in *Dividends of kinship: Meanings and uses of social relatedness.* Edited by Schweitzer, P. P., pp. 33–60. London: Routledge.

Nuttall, M. 2008. "Arsarnerit: Inuit and the heavenly game of football," in *Words for football: World literature on soccer and the human condition.* Edited by Raab, A., T. Satterleem, and J. Turnbull, pp. 274–282. Lincoln: University of Nebraska Press.

Oosten, J. G. 1986. Male and female in Inuit shamanism. *Etudes Inuit Studies* 10 (1–2): 115–131.

Oosten, J. G., and F. Laugrand. 2004. Time and space in the perception of non-human beings among the Inuit of Northeast Canada. *Jordens Folk: Journal of the Danish Ethnographic Society* 46–47: 85–120.

Oshana, M. 1998. Personal autonomy in society. *Journal of Social Philosophy* 29 (1): 89–102.

Oshana, M. 2006. *Personal autonomy in society.* Aldershot: Ashgate.

Ouellette, N. 2002. Les tuurngait dans le Nunavik occidental contemporain. *Études Inuit Studies* 26 (2): 107–131.

Paine, R. 1961. What is gossip about? An alternative hypothesis. *Man* 2 (2): 278–285.

Pálsson, G. 1994. Enskilment at sea. *Man* 29 (4): 901–925.

Pauktuutit. 2006. *Inuit ways: A guide to Inuit culture.* Pauktuutit, Inuit Women of Canada. www.pauktuutit.ca/pdf/publications/pauktuutit/InuitWay_e.pdf (10.03.2010).

Pedersen, B. K. 2009a. "Children and orality – self reported body and emotional experiences with horror stories," in *Proceedings of the 15th Inuit Studies Conference.* Edited by Beatrice, C. and M. Therrien. Paris: INALCO/CNR. www.inuitoralityconference. com/ (12.01.2010), pp. 1–12.

Pedersen, B. K. 2009b. "A narrative on narratives in contemporary Greenland," in *Proceedings of the 15th Inuit Studies Conference.* Edited by Beatrice, C. and M. Therrien. Paris: INALCO/CNR. www.inuitoralityconference.com/ (12.01.2010), pp. 1–14.

Pedersen, F. 2005. *Jagd auf den Qivittoq.* Wuppertal: Peter Hammer Verlag.

Perrot, M. 1975. L'émigration du Groenland oriental, un nouveau schème culturel. *Objets et Mondes* 15 (2): 169–176.

Petersen, H. C., and M. Hauser. 2006. *Kalaallit inngerutinil atuinerat. Trommesangtraditionen i Grønland. The drum song tradition in Greenland.* Nuuk: Forlaget Atuagkat.

Petersen, R. 1964. The Greenland Tupilak. *Folk* 6 (2): 73–94.

Petersen, R. 1975. On the East Greenlandic dialect in comparison with the West Greenlandic. *Objects et Mondes* 15 (2): 177–182.

Petersen, R. 1984a. "East Greenland after 1950," in *Handbook of North American Indians: 5 Arctic*. Edited by Damas, D., pp. 718–723. Washington: Smithsonian Institution.

Petersen, R. 1984b. "East Greenland before 1950," in *Handbook of North American Indians: 5 Arctic*. Edited by Damas, D., pp. 622–640. Washington: Smithsonian Institution.

Petersen, R. 1996. "Body and soul in ancient Greenlandic religion," in *Shamanism and northern ecology*. Edited by Pentikaien, J., pp. 67–78. Berlin: Mouton de Gruyter.

Petersen, R. 2003. *Settlements, kinship and hunting grounds in traditional Greenland: A comparative study of local experiences from Upernavik and Ammassalik*. Copenhagen: MoG Man & Society, Danish Polar Center.

Peterson, N. 1993. Demand sharing: Reciprocity and the pressure for generosity among foragers. *American Anthropologist* 4: 860–874.

Philips, S. U. 2007. "Participant structures and communicative competence: Warm Springs children in community and classroom," in *A cultural approach to interpersonal communication: Essential readings*. Edited by Monaghan, L. and J. E. Goodman, pp. 378–395. Malden, Oxford, Carlton: Blackwell Publishing.

Phillips Davison, W. 1958. The public opinion process. *Public Opinion Quarterly* 22: 91–106.

Price, J. A. 1975. Sharing: The integration of intimate economies. *Anthropologica* 17 (1): 3–27.

Price, V. 1989. Social identification and public opinion: Effects of communicating group conflict. *Public Opinion Quarterly* 53: 197–224.

Qúpersimân, G. 1982. *Min eskimoiske fortid: en østgrønlandsk åndemaners erindringer*. Edited by Otto Sandgreen, Nuuk: Grønlandske forlag.

Radcliffe-Brown, A. R. 1940. On joking relationships. *Africa: Journal of the International African Institute* 13 (3): 195–210.

Radcliffe-Brown, A. R. 1949. A further note on joking relationships. *Africa: Journal of the International African Institute* 19: 133–140.

Rasmussen, K. 1921. *Myter og sagn på Grønland. I. Østgrønlændere*. Copenhagen: Gylendalske Boghandel, Nordisk Forlag.

Rasmussen, K. 1938. *Knud Rasmussen's posthumous notes on the life and doings of the East Greenlanders in olden times*. Medellelser om Grønland. Copenhagen: C.A. Reitzel.

Rasmussen, S. 2008. Personhood, self, difference, and dialogue (commentary on Chaudhary). *International Journal for Dialogical Science* 3 (1): 31–54.

Riches, D. 1975. Cash, credit and gambling in a modern Eskimo economy: Speculations on origins of spheres of economic exchange. *Man* 10 (1): 21–33.

Riches, D. 2000. The holistic person; or, the ideology of egalitarianism. *The Journal of the Royal Anthropological Institute* 6: 669–685.

Riches, D. 2004. "Space-time, ethnicity, and the limits of Inuit and New Age egalitarianism," in *Property and equality, vol. 1: Ritualization, sharing, egalitarianism*. Edited by Widlok, T. and W. G. Tadesse, pp. 62–76. New York, Oxford: Berghahn Books.

Ridington, R. 1988. Knowledge, power, and the individual in subarctic hunting societies. *American Anthropologist* 90 (1): 98–110.

Rink, H. 1875. *Tales and traditions of the Eskimo with a sketch of their habits, religion, language and other peculiarities*. Edinburgh, London: William Blackwood and Sons.

Robbe, B. 1975a. Le traitement des peaux de phoque chez les Ammassalimiut observé en 1972 dans le village de Tîleqilaq. *Objets et Mondes* 15 (2): 199–208.

Robbe, P. 1975b. Partage du gibier chez les Ammassalimiut observé en 1972 dans le village Tileqilaq. *Objets et Mondes* 15 (2): 209–223.

Robbe, P. 1977. Orientation et repérage chez les Tileqilamiut (côte est du Groenland). *Études Inuit Studies* 1 (2): 73–83.

Robbe, P. 1981. Les noms de personnes chez les Ammassalimiut. *Études Inuit Studies* 5 (1): 45–82.

Robbe, P. 1983. Existence et mode d'intervention des sorciers (*ilisiitsut*) dans la société inuit d'Ammassalik. *Études Inuit Studies* 7 (1): 25–40.

Robbe, P. 1994. *Les inuit d'Ammassalik, chausseurs de l'Arctique*. Tome 156. Paris: Mémoires du Muséum National d'Histoire Naturelle.

Robbe, P., and L.-J. Dorais. 1986. *Tunumiit oraasiat. Tunumiut oqaasii. Det østgronldlandske sprog*. The East Greenlandic Inuit Language, La langue inuit du Groenland de l'Est: Collection Nordicana 49. Québec: Centre d'Études Nordiques, Université Laval.

Robert-Lamblin, J. 1981. Changement de sexe de certains enfants d'Ammassalik (Est Groenland): un rééquilibrage du sex ratio familia? *Études Inuit Studies* 5 (1): 117–126.

Robert-Lamblin, J. 1984. L'expression de la violence dans la société ammassalimiut (Côte orientale du Groenland). *Études Rurales* 95–96: 115–129.

Robert-Lamblin, J. 1986. *Ammassalik, East Greenland – end or persistence of an isolate? Anthropological and demographical study on change*. Meddelelser om Grønland, Man & Society 10. Copenhagen: Nyt Nordisk Forlag.

Robert-Lamblin, J. 1992. "Life, death and therapy among the East Greenlanders: Tradition and acculturation," in *Regard sur l'avenir/Looking to the future. Communications du 7e Congrès d'Études Inuit/Papers from the 7th Inuit studies Conference*. Edited by Dufour, M.-J. and F. Thérien, pp. 115–122. Québec: Université Laval.

Robert-Lamblin, J. 1996. Les dernières manifestation du chamanisme au Groenland orientale. *Boréales, Revue du Centre de Recherches Inter-Nordiques* 65–69: 115–130.

Robert-Lamblin, J. 1997a. "Death in traditional East Greenland: Age, causes and rituals: A contribution from anthropology to archaeology," in *Fifty years of Arctic research: Anthropological studies from Greenland to Siberia*. Ethnographical Series, Vol. 18. Edited by Gilberg, R. and H. C. Gulløv, pp. 261–268. National Museum of Denmark, Copenhagen: Publications of the National Museum.

Robert-Lamblin, J. 1997b. Les chamanes du Groenland oriental: éléments biographiques et généalogiques. *Études Inuit Studies* 21 (1–2): 269–292.

Robert-Lamblin, J. 1999a. Famille biologique et famille social à Ammassalik (cote Est du Groenland). Influence des changements recents sur la structure familiale et la fécondité des femmes. *Études Inuit Studies* 2 (2): 23–36.

Robert-Lamblin, J. 1999b. La famille, le village, la ville: dynamique du changement social au Groenland oriental de 1960 à 1990. *Études Inuit Studies* 23 (1–2): 35–53.

Robert-Lamblin, J. 2000. Cycles de vie et transitions dans la société est-groenlandaise: D'hier à aujourd'hui. *Études Inuit Studies* 24 (2): 47–64.

Roepstorff, A. 2007. "Navigating the brainscape: When knowing becomes seeing," in *Skilled visions: Between apprenticeship and standards*. Edited by Grasseni, C., pp. 191–206. Oxford: Berghahn.

Roepstorff, A., N. Bubandt, and K. Kull. Editors. 2004. *Imagining nature: Practices of cosmology and identity*. Aarhus: Aarhus University Press.

Romalis, S. 1983. The East Greenland Tupilaq image: Old and new visions. *Études Inuit Studies* 7 (1): 152–160.

Rønsager, M. 2002. *Udviklingen i grønlændernes sundheds- og sygdomsopfattelse 1800–1930*. SIFs Grønlandsskrifter nr. 14. Copenhagen: Statens Institut for Folkesundhed (SIF).

Rosaldo, M. Z. 1974. "Woman, culture, and society: A theoretical overview," in *Woman, culture, and society*. Edited by Rosaldo, M. Z. and L. Lamphere, pp. 17–42. Stanford: Stanford University Press.

Rosaldo, M. Z. 1980. *Knowledge and passion: Ilongot notions of self and social life*. Cambridge: Cambridge University Press.

Rosaldo, M. Z. 1984. "Towards an anthropology of self and feeling," in *Culture theory: Essays on mind, self, and emotions*. Edited by Shweder, R. A. and R. A. Levine, pp. 137–157. Cambridge: Cambridge University Press.

Rosaldo, M. Z., and L. Lamphere. Editors. 1974. *Woman, culture, and society*. Stanford: Stanfors University Press.

Rosing, J. 1957. Den østgrønlanske 'Maskekultur'. *Grønland – Den grønlandske Selskab* 7: 241–252.

Rosing, J. 1960. *îsímardik. Den store drabsmand*. Skjern: Det Grønlandske Selskabs Skrifter XX. Gullanders Bogtrykkeri.

Rosing, J. 1963. *Sagn og saga fra Ammassalik*. Copenhagen: Rhodos.

Rosing, J. 1969. Sangkamp hos Angmagssalik-folket. *Tidskrift for Grønlands Retsvæsen* 3: 70–89.

Rosing, J. 1993. *Hvis i vågner til havbik, En slægtssaga fra Østgrønland*. Copenhagen: Borgens Forlag.

Rundstrom, A. R. 1990. A cultural interpretation of Inuit map accuracy. *Geographical Review* 80 (1): 155–168.

Sahlins, M. D. 1972. *Stone age economies*. London: Travistock.

Saladin d'Anglure, B. 1986. Du foetus au chamane: la construction d'un 'troisième sexe' inuit. *Études Inuit Studies* 10 (1–2): 25–114.

Sandgreen, O. 1987. *Øje for øje of tand for tand*. Nuuk: Otto Sandgreen Forlag.

Saville-Troike, M. 2003. *The ethnography of communication: An introduction*. Language in Society. Oxford: Blackwell.

Schieffelin, B. B. 1986. "Teasing and shaming in Kaluli children's interactions," in *Language socialization across cultures*. Studies in the Social and Cultural Foundation of Language, Vol. 3. Edited by Schieffelin, B. B. and E. Ochs, pp. 165–181. Cambridge, New York, Melbourne: Cambridge University Press.

Schiffrin, D. 1993. "'Speaking for another' in sociolinguistic interviews: Alignment, identities, and frames," in *Framing in discourse*. Edited by Tannen, D., pp. 231–262. New York, Oxford: Oxford University Press.

Schoeman, F. D. 1984. *Philosophical dimensions of privacy: An anthology*. Cambridge: Cambridge University Press.

Schoeman, F. D. 1994. "Gossip and privacy," in *Good gossip*. Edited by Goodman, R. F. and A. Ben-Ze'ev, pp. 72–82. Lawrence, KS: University Press of Kansas.

Schottman, W. 1993. Proverbial dog names of the Baatombu: A strategic alternative to silence. *Language in Society* 22 (4): 539–554.

Scott, J. 1985. *Weapons of the weak: Everyday forms of peasant resistance*. New Haven, London: Yale University Press.

Scott, J. 1990. *Domination and the arts of resistance: Hidden transcripts*. New Haven, London: Yale University Press.

Searle, J. 1969. *Speech acts: An essay in the philosophy of language*. Cambridge: Cambridge University Press.

Seitz, A., and S. Elixhauser. 2008. "Sermiligaaq 65°54'N, 36°22'W." Documentary film, 63 min, DigiBeta, HDV 16:9, Munich.

Sejersen, F. 2002. *Local knowledge, sustainability and visionscapes in Greenland.* Copenhagen: Eskimologis Skrifter, nr. 17, Department of Eskimology, University of Copenhagen.

Shannon, C. E., and W. Weaver. 1964. *The mathematical theory of communication.* Urbana: The University of Illinois Press.

Sheets-Johnstone, M. 1999. *The primacy of movement.* Advances in Consciousness Research, 1381–589X. Amsterdam: J. Benjamins.

Sheets-Johnstone, M. 2000. Kinetic tactile-kinesthetic bodies: Ontogenetical foundations of apprenticeship learning. *Human Studies* 23 (4): 343–370.

Sheets-Johnstone, M. 2009. *The corporeal turn: An interdisciplinary reader.* Exeter: Imprint Academic.

Sherzer, J. 1983. *Kuna ways of speaking: An ethnographic perspective.* Austin: University of Texas Press.

Sherzer, J. 1990. *Verbal art in San Blas: Kuna culture through its discourse.* Cambridge: Cambridge University Press.

Simmel, G. 1969. "Sociology of the senses: Visual interaction," in *Introduction to the science of sociology.* 3rd edition. Edited by Burgesss, R. E. P. A. E. W., pp. 146–150. Chicago: University of Chicago Press.

Smidt, C. M., and I. M. Smidt. 1975. Du chant au Tambour aux cours de justice locales. *Objets et Mondes* 15 (2): 243–246.

Sonne, B. 1982. The ideology and practice of blood feuds in East and West Greenland. *Études Inuit Studies* 6 (2): 21–50.

Sonne, B. 1986. Toornaarsuk, an historical proteus. *Arctic Anthropology* 23 (1–2): 199–219.

Sonne, B. 1990. "The acculturative role of sea woman: Early contact relations between Inuit and Whites as revealed in the origin myth of sea woman." *Medellelser om Grønland (Man & Society)* 13: 1–34.

Sonne, B. 1996. "Genuine humans and 'others': Criteria of 'otherness' at the beginning of colonization in Greenland," in *Cultural and social research in Greenland 95/96.* Edited by Ilisimatusarfik, pp. 241–252. Nuuk: Ilisimatusarfik/Atuakkiorfik.

Sonne, B. 2003. "Er tavshed guld? Hemmeligholdelse i Østgrønland før – og endnu?," in *Grønlandsk Kultur- og Samfunds Forskning 2003.* Edited by Ilisimatusarfik, pp. 203–220. Nuuk: Forlaget Atuagkat.

Sowa, F. 2014. *Indigene Völker in der Weltgesellschaft: die kulturelle Identität der grönländischen Inuit im Spannungsfeld von Natur und Kultur.* Bielefeld: Transcript.

Stairs, A. 1992. Self-image, world-image: Speculations on identity from experiences with Inuit. *Ethos* 20: 116–126.

Statistics Greenland. 2017. *Greenland in figures 2017.* 14th revised edition. Edited by Vahl, B. and N. Kleemann, Statistics Greenland. Government of Greenland. www.stat.gl/publ/kl/GF/2017/pdf/Greenland%20in%20Figures%202017.pdf (24.05.2017).

Stern, P. 2003. Upside-down and backwards: Time discipline in a Canadian Inuit Town. *Anthropologica* 45: 147–161.

Stevenson, L. 2014. *Life beside itself: Imagining care in the Canadian Arctic.* Berkeley: University of California Press.

Stewart, P. J., and A. Strathern. 2004. *Witchcraft, sorcery, rumors, and gossip.* New Departures in Anthropology. Cambridge, New York: Cambridge University Press.

Strathern, M. 1990. *The gender of the gift.* Berkeley: University of California Press.

Strecker, I. 1993. Cultural variations in the concept of 'face'. *Multilingua* 12: 119–141.

Strecker, I. 2004. "To share or not to share: Notes about authority and anarchy among the Hamar of Southern Ethiopia," in *Property and equality, vol. 1: Ritualization, sharing, egalitarianism*. Edited by Widlok, T. and W. G. Tadesse, pp. 175–189. New York, Oxford: Berghahn Books.

Stuckenberger, A. N. 2009. "Of what a house can do," in *Proceedings of the 15th Inuit Studies Conference*. Edited by Beatrice, C. and M. Therrien. Paris: INALCO/CNR. www. inuitoralityconference.com/ (12.01.2009), pp. 1–13.

Suchman, L. A. 1987. *Plans and situated actions: The problem of human-machine communication*. Cambridge, New York: Cambridge University Press.

Sugawara, K. 2004. "Possession, equality and gender relations in |Gui discourse," in *Property and equality, vol. 1: Ritualization, sharing, egalitarianism*. Edited by Widlok, T. and W. G. Tadesse, pp. 105–129. New York, Oxford: Berghahn Books.

Tambiah, S. J. 1968. The magical power of words. *Man* 3: 175–208.

Tannen, D. 2007. *Talking voices: Repetition, dialogue, and imagery in conversational discourse*. 2nd edition. Cambridge: Cambridge University Press.

Tannen, D., and M. Saville-Troike. Editors. 1985. *Perspectives on silence*. Norwood, NJ: Ablex Publishing Corporation.

Tersis, N. 2008. *Forme et sens des mots du Tunumiisut. Lexique inuit du Groenland oriental*. SELAF No. 445. Louvain, Paris, Dudley, MA: Peeters.

Thalbitzer, W. 1914a. The Ammassalik Eskimo: Contributions to the ethnology of the East Greenland natives. *Meddelelser om Grønland* 39 (part one): 1–755.

Thalbitzer, W. 1914b. "The East Greenlandic dialect: According to the annotations made by the Danish east coast expedition to Kleinschmidt's Greenlandic dictionary by H. Rink 1887," in *The Ammassalik Eskimo: Contributions to the ethnology of the East Greenland natives*. Vol. 39 (1). Edited by Thalbitzer, W., pp. 205–223. Copenhagen: Meddelelser om Grønland.

Thalbitzer, W. 1914c. "Ethnographical collections from East Greenland (Angmagsalik and Nualik) made by G. Holm, G. Amdrup and J. Petersen and described by W. Thalbitzer," in *The Ammassalik Eskimo: Contributions to the ethnology of the East Greenland natives*. Vol. 39 (7). Edited by Thalbitzer, W., pp. 321–732. Copenhagen: Meddelelser om Grønland.

Thalbitzer, W. 1921. The Ammassalik Eskimo: Language and folklore. *Meddelelser om Grønland* 40 (3, part two): 450.

Thalbitzer, W. 1930. Les magiciens esquimaux. Leurs conceptions du monde, de l'âme et de la vie. *Journal de la Société des Américanistes de Paris* 22: 73–106.

Thalbitzer, W. 1941. The Ammassalik Eskimo: Contributions to the ethnology of the East Greenland natives: Nr. 4 Social customs and mutual aid. *Meddelelser om Grønland* 40 (4, part two, second half-volume): 569–739.

Thalbitzer, W. 1974. *A phonetical study of the Eskimo language*. 1st edition 1904. New York: AMS Press.

Therrien, M. 1987. *Le corps inuit: Québec arctique*. Collection Arctique 1. Paris: SELAF/ PUB.

Therrien, M. 1997. "Inuit concepts and notions regarding the Canadian justice system," in *Legal glossary/Glossaire juridique*. Edited by Brice-Bennet, D., pp. 270–275. Iqaluit: Nunavut Arctic College.

Therrien, M. 2002. Rêves d'une apprentie chamane inuit. *Cahiers De Littérature Orale* 51: 169–183.

Therrien, M. 2008. "Tout révéler et rester discret chez les Inuit de l'Arctique oriental canadien," in *Paroles interdites*. Edited by Therrien, M., pp. 251–285. Paris: Karthala.

Therrien, M., and B. Collignon. 2009. "Introduction to Inuit orality," in *Proceedings of the 15th International Inuit Conference*. Edited by Therrien, M. and B. Collignon. www. inuitoralityconference.com/index_eng.html (12.01.2009).

Therrien, M., and T. Qumaq. 1995. Corps sains, corps malade chez les Inuit. Une tension entre l'intérieur et l'extérieur. Entretiens avec Taamusi Qumaq. *Recherches amérindiennes au Québec* 25 (12): 71–84.

Thorslund, J. 1992. "Why do they do it? Proposal for a theory of Inuit suicide," in *Regard sur l'avenir/Looking to the future. Communications du 7e Congrès d'Études Inuit/ Papers from the 7th Inuit studies Conference*. Edited by Dufour, M.-J. and F. Thérien, pp. 149–161. Québec: Université Laval.

Tilley, C. 1994. *A phenomenology of landscape: Places, paths and monuments*. Oxford: Berg.

Tilley, C. 2004. *The materiality of stone*. Oxford: Berg.

Turnbull, D. 2002. Performance and narrative, bodies and movement in the construction of places and objects, spaces and knowledges: The case of the Maltese megaliths. *Theory, Culture & Society* 19 (5): 125–143.

Tylor, E. B. 1913 [1871]. *Primitive culture*. 2 vols. London: John Murray.

van Dommelen, P. 1999. "Exploring everyday places and cosmologies," in *Archaeologies of landscape: Contemporary perspectives*. Edited by Ashmore, W. and A. B. Knapp, pp. 277–285. Oxford: Blackwell Publishers.

Victor, P.-É., C. Enel, and E. Maqe. 1991. *Chants d'Ammassalik*. Meddeleser im Grønland (Man & Society), 286 pp. Copenhagen: Danish Polar Centre.

Victor, P.-É., and J. Robert-Lamblin. 1989. *La civilisation du phoque 1. Jeux, gestes et techniques des Eskimo d'Ammassalik*. Paris: Éditions Raymond Chabaud/Armand Colin.

Victor, P.-É., and J. Robert-Lamblin. 1993. *La civilisation du phoque 2. Légendes, rites et croyances des Eskimo d'Ammassalik*. Paris: Éditions Raymond Chabaud/Armand Colin.

Viveiros de Castro, E. 1998. Cosmological deixis and amerindian perspectivism. *Journal of the Royal Anthropological Institute* 4 (3): 469–488.

Vom Bruck, G. 1997. A house turned inside out: Inhabiting space in a Yemeni city. *Journal of Material Culture* 2 (2): 139–172.

Wachowich, N., in collaboration with Apphia Agalakti Awa, Rhoda Kaukjak Katsak, and Sandra Pikujak Katsak. 2001. *Saqiying: Stories from the lives of three Inuit women*. McGill-Queen's Native and Northern Series. Montreal, Kingston: McGill-Queen's University Press.

Wagner Sørensen, B. 1994. *Magt eller afmagt? Køn, følelser og vold i Grønland*. Århus: Akademisk Forlag.

Wagner Sørensen, B. 2001. 'Men in transition': The representation of men's violence against women in Greenland. *Violence against Women* 7 (7): 826–847.

Watzlawick, P., J. B. Bavelas, and D. D. Jackson. 1967. *Pragmatics of human communication: A study of interactional patterns, pathologies, and paradoxes*. New York: W. W. Norton.

Wenzel, G. 1991. *Animal rights, human rights: Ecology, economy and ideology in the Canadian Arctic*. Toronto: University of Toronto Press.

Westlund, A. C. 2008. Rethinking relational autonomy: University of Wisconsin-Milwaukee, 14 pp. https://pantherfile.uwm.edu/westlund/www/rra-short-posted04-23-08.pdf (04.12.2009).

Whitridge, P. 2004. Landscape, houses, bodies, things: 'Place' and the archaeology of Inuit imagineries. *Journal of Archaeological Method and Theory* 11 (2).

Wikan, U. 1993. "Beyond the words: The power of resonance," in *Beyond boundaries: Understanding, translation, and anthropological discourse*. Explorations in Anthropology. Edited by Pálsson, G., pp. 184–209. Oxford: Berg.

Willerslev, R. 2007. *Soul hunters: Hunting, animism and personhood among the Siberian Yukaghir*. Berkeley: University of California Press.

Willerslev, R. 2012. Laughing at the spirits in North Siberia: Is animism being taken too seriously? *e-flux* 36 (7): 13–22.

Williams, B. 1987. Humor, linguistic ambiguity, and disputing in a Guyanese community. *International Journal of the Sociology of Language* 65: 79–94.

Wilson, P. 1974. Filcher of good names: An enquiry into anthropology and gossip. *Man* 9 (1): 93–103.

Wittgenstein, L. 1953/2001. *Philosophische Untersuchungen/Philosophical investigations*. Oxford: Blackwell Publishing.

Woodburn, J. 1982. Egalitarian societies. *Man* 17 (3): 431–451.

Woodburn, J. 1998. "Sharing is not a form of exchange: An analysis of property-sharing in immediate return hunter-gatherer societies," in *Property relations: Renewing the anthropological tradition*. Edited by Hann, C., pp. 48–63. Cambridge: Cambridge University Press.

Zerubavel, E. 2006. *The elephant in the room: Silence and denial in everyday life*. Oxford, New York: Oxford University Press.

Index